Calculus of Variations and
Optimal Control Theory

Calculus of Variations and Optimal Control Theory

A Concise Introduction

Daniel Liberzon

PRINCETON UNIVERSITY PRESS

PRINCETON AND OXFORD

Published by Princeton University Press, 41 William Street, Princeton, New Jersey
08540

In the United Kingdom: Princeton University Press, 6 Oxford Street, Woodstock,
Oxfordshire OX20 1TW

ISBN: 978-0-691-15187-8

Library of Congress Control Number: 2011935625

British Library Cataloging-in-Publication Data is available

This book has been composed in L^AT_EX

The publisher would like to acknowledge the author of this volume for providing the
digital files from which this book was printed

press.princeton.edu

10 9 8 7 6 5 4 3 2 1

Since the building of the universe is perfect and is created by the wisdom creator, nothing arises in the universe in which one cannot see the sense of some maximum or minimum.

—Leonhard Euler

The words "control theory" are, of course, of recent origin, but the subject itself is much older, since it contains the classical calculus of variations as a special case, and the first calculus of variations problems go back to classical Greece.

—Hector J. Sussmann

Contents

Preface

This book grew out of my lecture notes for a graduate course on optimal control theory which I taught at the University of Illinois at Urbana-Champaign during the period from 2005 to 2010. While preparing the lectures, I have accumulated an entire shelf of textbooks on calculus of variations and optimal control systems. Although some of them are excellent, navigating and combining the material from so many sources was a daunting task, and I was unable to find a single text with all the features that I wanted:

APPROPRIATE PRESENTATION LEVEL. I needed a friendly introductory text accessible to graduate students who have not had extensive training in control theory or mathematical analysis; a text that is rigorous and builds a solid understanding of the subject without getting bogged down in technicalities.

LOGICAL AND NOTATIONAL CONSISTENCY AMONG TOPICS. There are intimate connections between the classical calculus of variations, the maximum principle, and the Hamilton-Jacobi-Bellman theory, which I wanted to emphasize throughout the course. Due to differences in notation and presentation style in the existing sources, it is very difficult for students (and even for instructors) to piece this material together.

PROOF OF THE MAXIMUM PRINCIPLE. The maximum principle is a centerpiece of optimal control theory which also elucidates and puts in perspective earlier developments in calculus of variations, and I felt it was important to cover its proof. This is a challenging task because a complete proof is rather long and, without adequate preparation and guidance, the effort required to understand it can seem prohibitive to most students.

HISTORICAL PERSPECTIVE. A course on optimal control gives students a unique chance to get exposed to the work of some of the greatest mathematical minds and trace the beautiful historical development of the subject, from early calculus of variations results to modern optimal control theory. A good text, while of course focusing on the mathematical developments, should also give students a sense of the history behind them.[1]

---

[1] In Tom Stoppard's famous play *Arcadia* there is an interesting argument between a mathematician and a historian about what is more important: scientific progress or the

MANAGEABLE SIZE. It is very useful to have a text compact enough so that all of the material in it can be covered in one semester, since very few students will take more than one course specifically on this subject.

The present book is an attempt to meet all of the above challenges (inevitably, with varying degrees of success). The comment about proving the maximum principle in class deserves some elaboration. I believe that to build the understanding necessary to correctly and effectively use the maximum principle, it is very helpful—if not essential—to master its proof. In my experience, the proof given in this book is not too difficult for students to follow. This has to do not just with the clarity and structure of the proof itself, but also with the fact that special attention is paid to preparing students for this proof earlier on in the course (as well as helping them digest it through subsequent discussions). On the other hand, due to the central role that the proof of the maximum principle plays in this book, instructors not planning to cover this proof in class will not fully benefit from adopting this text.

While some prior exposure to control theory, mathematical analysis, and optimization is certainly helpful, this book is sufficiently self-contained so that any motivated graduate student specializing in a technical subject (such as engineering or mathematics) should be able to follow it. Depending on the student's background, some supplementary reading may be necessary from time to time; the notes and references located at the end of each chapter should facilitate this process.

Chapters 1–6 form the main part of the book and can serve as the basis for a one-semester course. Depending on the pace, the instructor may also have time for some or all of the advanced topics discussed in Chapter 7. In this last chapter I included only topics that in my opinion directly extend and enhance the understanding of the core material. The instructor may give preference to some other areas instead, such as the important issue of state constraints in optimal control, the classical subject of stochastic optimal control and estimation, the very popular model predictive control, or the numerous applications of optimal control theory. Fortunately, good references covering these topics are readily available. It is also possible that the instructor will want to further elaborate on some aspects of the theory presented in Chapters 1–6; in this regard, the end-of-chapter notes and references may be a useful resource.

About 60 exercises are interspersed throughout the first six chapters and represent an integral part of the book. My intention is for students to start working on each exercise as soon as the corresponding material is covered. A solutions manual is available upon request for instructors.

---

personalities behind it. In a technical book such as this one, the emphasis is clear, but some flavor of the history can enrich one's understanding and appreciation of the subject.

I am grateful to my colleagues Tamer Başar, Yi Ma, and Bill Perkins who shared their valuable experiences and lecture notes from their offerings of the optimal control course at UIUC. I am also indebted to students and friends who suffered through earlier drafts of this book and provided numerous corrections and suggestions, along with stimulating discussions and much-needed encouragement; I would like to especially thank Sairaj Dhople, Michael Margaliot, Nathan Shemonski, Hyungbo Shim, Guosong Yang, and Jingjin Yu. I appreciated the helpful and thoughtful comments of the reviewers and the support and professionalism of Vickie Kearn and other editorial staff at Princeton University Press. I thank Glenda Krupa for the careful and thorough copyediting. On the artistic side, I thank Polina Ben-Sira for the very cool cover drawing, and my wife Olga not only for moral support but also for substantial help with all the figures inside the book. Finally, I am thankful to my daughter Ada who told me that I should write the book and gave me a toy tangerine which helped in the process, as well as to my little son Eduard whose arrival definitely boosted my efforts to finish the book.

I decided not to eliminate all errors from the book, and instead left several of them on purpose in undisclosed locations as a way to provide additional educational experience. To report success stories of correcting errors, please contact me at my current email address (which is easy to find on the Web).

Daniel Liberzon

Urbana, IL

Calculus of Variations and
Optimal Control Theory

Chapter One

Introduction

1.1 OPTIMAL CONTROL PROBLEM

We begin by describing, very informally and in general terms, the class of optimal control problems that we want to eventually be able to solve. The goal of this brief motivational discussion is to fix the basic concepts and terminology without worrying about technical details.

The first basic ingredient of an optimal control problem is a *control system*. It generates possible behaviors. In this book, control systems will be described by ordinary differential equations (ODEs) of the form

$$\dot{x} = f(t, x, u), \qquad x(t_0) = x_0 \tag{1.1}$$

where x is the *state* taking values in $\mathbb{R}^n$, u is the *control input* taking values in some *control set* $U \subset \mathbb{R}^m$, t is *time*, t_0 is the *initial time*, and x_0 is the *initial state*. Both x and u are functions of t, but we will often suppress their time arguments.

The second basic ingredient is the *cost functional*. It associates a cost with each possible behavior. For a given initial data (t_0, x_0), the behaviors are parameterized by control functions u. Thus, the cost functional assigns a cost value to each admissible control. In this book, cost functionals will be denoted by J and will be of the form

$$J(u) := \int_{t_0}^{t_f} L(t, x(t), u(t)) dt + K(t_f, x_f) \tag{1.2}$$

where L and K are given functions (*running cost* and *terminal cost*, respectively), t_f is the *final* (or *terminal*) *time* which is either free or fixed, and $x_f := x(t_f)$ is the *final* (or *terminal*) *state* which is either free or fixed or belongs to some given target set. Note again that u itself is a function of time; this is why we say that J is a *functional* (a real-valued function on a space of functions).

The optimal control problem can then be posed as follows: Find a control u that minimizes $J(u)$ over all admissible controls (or at least over nearby controls). Later we will need to come back to this problem formulation and fill in some technical details. In particular, we will need to specify what regularity properties should be imposed on the function f and on the admissible controls u to ensure that state trajectories of the control system are well defined. Several versions of the above problem (depending, for example, on the role of the final time and the final state) will be stated more precisely when we are ready to study them. The reader who wishes to preview this material can find it in Section 3.3.

It can be argued that optimality is a universal principle of life, in the sense that many—if not most—processes in nature are governed by solutions to some optimization problems (although we may never know exactly what is being optimized). We will soon see that fundamental laws of mechanics can be cast in an optimization context. From an engineering point of view, optimality provides a very useful design principle, and the cost to be minimized (or the profit to be maximized) is often naturally contained in the problem itself. Some examples of optimal control problems arising in applications include the following:

- Send a rocket to the moon with minimal fuel consumption.

- Produce a given amount of chemical in minimal time and/or with minimal amount of catalyst used (or maximize the amount produced in given time).

- Bring sales of a new product to a desired level while minimizing the amount of money spent on the advertising campaign.

- Maximize throughput or accuracy of information transmission over a communication channel with a given bandwidth/capacity.

The reader will easily think of other examples. Several specific optimal control problems will be examined in detail later in the book. We briefly discuss one simple example here to better illustrate the general problem formulation.

Example 1.1   *Consider a simple model of a car moving on a horizontal line. Let $x \in \mathbb{R}$ be the car's position and let u be the acceleration which acts as the control input. We put a bound on the maximal allowable acceleration by letting the control set U be the bounded interval $[-1, 1]$ (negative acceleration corresponds to braking). The dynamics of the car are $\ddot{x} = u$. In order to arrive at a first-order differential equation model of the form (1.1), let us relabel the car's position x as x_1 and denote its velocity $\dot{x}$ by x_2. This gives the control system $\dot{x}_1 = x_2$, $\dot{x}_2 = u$ with state $\begin{pmatrix} x_1 \\ x_2 \end{pmatrix} \in \mathbb{R}^2$. Now, suppose that we want to "park" the car at the origin, i.e., bring it to rest there, in minimal time. This objective is captured by the cost functional (1.2) with the constant running cost $L \equiv 1$, no terminal cost ($K \equiv 0$), and the fixed final state $\begin{pmatrix} 0 \\ 0 \end{pmatrix}$. We will solve this optimal control problem in Section 4.4.1.*

(The basic form of the optimal control strategy may be intuitively obvious, but obtaining a complete description of the optimal control requires some work.)   □

In this book we focus on the *mathematical theory* of optimal control. We will not undertake an in-depth study of any of the applications mentioned above. Instead, we will concentrate on the fundamental aspects common to all of them. After finishing this book, the reader familiar with a specific application domain should have no difficulty reading papers that deal with applications of optimal control theory to that domain, and will be prepared to think creatively about new ways of applying the theory.

We can view the optimal control problem as that of choosing the best *path* among all paths feasible for the system, with respect to the given cost function. In this sense, the problem is *infinite-dimensional*, because the space of paths is an infinite-dimensional function space. This problem is also a *dynamic* optimization problem, in the sense that it involves a dynamical system and time. However, to gain appreciation for this problem, it will be useful to first recall some basic facts about the more standard static finite-dimensional optimization problem, concerned with finding a minimum of a given function $f : \mathbb{R}^n \to \mathbb{R}$. Then, when we get back to infinite-dimensional optimization, we will more clearly see the similarities but also the differences.

The subject studied in this book has a rich and beautiful history; the topics are ordered in such a way as to allow us to trace its chronological development. In particular, we will start with *calculus of variations*, which deals with path optimization but not in the setting of control systems. The optimization problems treated by calculus of variations are infinite-dimensional but not dynamic. We will then make a transition to optimal control theory and develop a truly dynamic framework. This modern treatment is based on two key developments, initially independent but ultimately closely related and complementary to each other: the maximum principle and the principle of dynamic programming.

1.2   SOME BACKGROUND ON FINITE-DIMENSIONAL OPTIMIZATION

Consider a function $f : \mathbb{R}^n \to \mathbb{R}$. Let D be some subset of $\mathbb{R}^n$, which could be the entire $\mathbb{R}^n$. We denote by $|\cdot|$ the standard Euclidean norm on $\mathbb{R}^n$.

A point $x^* \in D$ is a *local minimum* of f over D if there exists an $\varepsilon > 0$ such that for all $x \in D$ satisfying $|x - x^*| < \varepsilon$ we have

$$f(x^*) \leq f(x). \tag{1.3}$$

In other words, x^* is a local minimum if in some ball around it, f does not attain a value smaller than $f(x^*)$. Note that this refers only to points in D;

the behavior of f outside D is irrelevant, and in fact we could have taken the domain of f to be D rather than $\mathbb{R}^n$.

If the inequality in (1.3) is strict for $x \neq x^*$, then we have a *strict* local minimum. If (1.3) holds for *all* $x \in D$, then the minimum is *global* over D. By default, when we say "a minimum" we mean a local minimum. Obviously, a minimum need not be unique unless it is both strict and global.

The notions of a (local, strict, global) *maximum* are defined similarly. If a point is either a maximum or a minimum, it is called an *extremum*. Observe that maxima of f are minima of $-f$, so there is no need to develop separate results for both. We focus on the minima, i.e., we view f as a *cost* function to be minimized (rather than a profit to be maximized).

1.2.1 Unconstrained optimization

The term "unconstrained optimization" usually refers to the situation where all points x sufficiently near x^* in $\mathbb{R}^n$ are in D, i.e., x^* belongs to D together with some $\mathbb{R}^n$-neighborhood. The simplest case is when $D = \mathbb{R}^n$, which is sometimes called the *completely unconstrained* case. However, as far as *local* minimization is concerned, it is enough to assume that x^* is an interior point of D. This is automatically true if D is an open subset of $\mathbb{R}^n$.

FIRST-ORDER NECESSARY CONDITION FOR OPTIMALITY

Suppose that f is a $\mathcal{C}^1$ (continuously differentiable) function and x^* is its local minimum. Pick an arbitrary vector $d \in \mathbb{R}^n$. Since we are in the unconstrained case, moving away from x^* in the direction of d or $-d$ cannot immediately take us outside D. In other words, we have $x^* + \alpha d \in D$ for all $\alpha \in \mathbb{R}$ close enough to 0.

For a fixed d, we can consider $f(x^* + \alpha d)$ as a function of the real parameter α, whose domain is some interval containing 0. Let us call this new function g:

$$g(\alpha) := f(x^* + \alpha d). \tag{1.4}$$

Since x^* is a minimum of f, it is clear that 0 is a minimum of g. Passing from f to g is useful because g is a function of a scalar variable and so its minima can be studied using ordinary calculus. In particular, we can write down the first-order Taylor expansion for g around $\alpha = 0$:

$$g(\alpha) = g(0) + g'(0)\alpha + o(\alpha) \tag{1.5}$$

where $o(\alpha)$ represents "higher-order terms" which go to 0 faster than α as α approaches 0, i.e.,

$$\lim_{\alpha \to 0} \frac{o(\alpha)}{\alpha} = 0. \tag{1.6}$$

We claim that

$$g'(0) = 0. \tag{1.7}$$

To show this, suppose that $g'(0) \neq 0$. Then, in view of (1.6), there exists an $\varepsilon > 0$ small enough so that for all nonzero α with $|\alpha| < \varepsilon$, the absolute value of the fraction in (1.6) is less than $|g'(0)|$. We can write this as

$$|\alpha| < \varepsilon, \ \alpha \neq 0 \quad \Rightarrow \quad |o(\alpha)| < |g'(0)\alpha|.$$

For these values of α, (1.5) gives

$$g(\alpha) - g(0) < g'(0)\alpha + |g'(0)\alpha|. \tag{1.8}$$

If we further restrict α to have the opposite sign to $g'(0)$, then the right-hand side of (1.8) becomes 0 and we obtain $g(\alpha) - g(0) < 0$. But this contradicts the fact that g has a minimum at 0. We have thus shown that (1.7) is indeed true.

We now need to re-express this result in terms of the original function f. A simple application of the chain rule from vector calculus yields the formula

$$g'(\alpha) = \nabla f(x^* + \alpha d) \cdot d \tag{1.9}$$

where

$$\nabla f := (f_{x_1}, \dots, f_{x_n})^T$$

is the *gradient* of f and $\cdot$ denotes inner product.[1] Whenever there is no danger of confusion, we use subscripts as a shorthand notation for partial derivatives: $f_{x_i} := \partial f / \partial x_i$. Setting $\alpha = 0$ in (1.9), we have

$$g'(0) = \nabla f(x^*) \cdot d \tag{1.10}$$

and this equals 0 by (1.7). Since d was arbitrary, we conclude that

$$\boxed{\nabla f(x^*) = 0} \tag{1.11}$$

This is the **first-order necessary condition for optimality**.

A point x^* satisfying this condition is called a *stationary point*. The condition is "first-order" because it is derived using the first-order expansion (1.5). We emphasize that the result is valid when $f \in \mathcal{C}^1$ and x^* is an interior point of D.

---

[1]There is no consensus in the literature whether the gradient is a column vector or a row vector. Treating it as a row vector would simplify the notation since it often appears in a product with another vector. Geometrically, however, it plays the role of a regular column vector, and for consistency we follow this latter convention everywhere.

SECOND-ORDER CONDITIONS FOR OPTIMALITY

We now derive another necessary condition and also a sufficient condition for optimality, under the stronger hypothesis that f is a $\mathcal{C}^2$ function (twice continuously differentiable).

First, we assume as before that x^* is a local minimum and derive a necessary condition. For an arbitrary fixed $d \in \mathbb{R}^n$, let us consider a Taylor expansion of $g(\alpha) = f(x^* + \alpha d)$ again, but this time include *second-order terms*:

$$g(\alpha) = g(0) + g'(0)\alpha + \frac{1}{2}g''(0)\alpha^2 + o(\alpha^2) \tag{1.12}$$

where

$$\lim_{\alpha \to 0} \frac{o(\alpha^2)}{\alpha^2} = 0. \tag{1.13}$$

We know from the derivation of the first-order necessary condition that $g'(0)$ must be 0. We claim that

$$g''(0) \geq 0. \tag{1.14}$$

Indeed, suppose that $g''(0) < 0$. By (1.13), there exists an $\varepsilon > 0$ such that

$$|\alpha| < \varepsilon, \ \alpha \neq 0 \quad \Rightarrow \quad |o(\alpha^2)| < \frac{1}{2}|g''(0)|\alpha^2.$$

For these values of α, (1.12) reduces to $g(\alpha) - g(0) < 0$, contradicting that fact that 0 is a minimum of g. Therefore, (1.14) must hold.

What does this result imply about the original function f? To see what $g''(0)$ is in terms of f, we need to differentiate the formula (1.9). The reader may find it helpful to first rewrite (1.9) more explicitly as

$$g'(\alpha) = \sum_{i=1}^{n} f_{x_i}(x^* + \alpha d)d_i.$$

Differentiating both sides with respect to α, we have

$$g''(\alpha) = \sum_{i,j=1}^{n} f_{x_i x_j}(x^* + \alpha d)d_i d_j$$

where double subscripts are used to denote second-order partial derivatives. For $\alpha = 0$ this gives

$$g''(0) = \sum_{i,j=1}^{n} f_{x_i x_j}(x^*)d_i d_j$$

or, in matrix notation,

$$g''(0) = d^T \nabla^2 f(x^*)d \tag{1.15}$$

where

$$\nabla^2 f := \begin{pmatrix} f_{x_1 x_1} & \cdots & f_{x_1 x_n} \\ \vdots & \ddots & \vdots \\ f_{x_n x_1} & \cdots & f_{x_n x_n} \end{pmatrix}$$

is the *Hessian* matrix of f. In view of (1.14), (1.15), and the fact that d was arbitrary, we conclude that the matrix $\nabla^2 f(x^*)$ must be positive semidefinite:

$$\boxed{\nabla^2 f(x^*) \geq 0} \quad \text{(positive semidefinite)}$$

This is the **second-order necessary condition for optimality**.

Like the previous first-order necessary condition, this second-order condition only applies to the unconstrained case. But, unlike the first-order condition, it requires f to be $\mathcal{C}^2$ and not just $\mathcal{C}^1$. Another difference with the first-order condition is that the second-order condition distinguishes minima from maxima: at a local maximum, the Hessian must be *negative* semidefinite, while the first-order condition applies to any extremum (a minimum or a maximum).

Strengthening the second-order necessary condition and combining it with the first-order necessary condition, we can obtain the following **second-order sufficient condition for optimality**: *If a $\mathcal{C}^2$ function f satisfies*

$$\nabla f(x^*) = 0 \quad \text{and} \quad \nabla^2 f(x^*) > 0 \quad \text{(positive *definite*)} \tag{1.16}$$

on an interior point x^* of its domain, then x^* is a strict local minimum of f. To see why this is true, take an arbitrary $d \in \mathbb{R}^n$ and consider again the second-order expansion (1.12) for $g(\alpha) = f(x^* + \alpha d)$. We know that $g'(0)$ is given by (1.10), thus it is 0 because $\nabla f(x^*) = 0$. Next, $g''(0)$ is given by (1.15), and so we have

$$f(x^* + \alpha d) = f(x^*) + \frac{1}{2} d^T \nabla^2 f(x^*) d \alpha^2 + o(\alpha^2). \tag{1.17}$$

The intuition is that since the Hessian $\nabla^2 f(x^*)$ is a positive definite matrix, the second-order term dominates the higher-order term $o(\alpha^2)$. To establish this fact rigorously, note that by the definition of $o(\alpha^2)$ we can pick an $\varepsilon > 0$ small enough so that

$$|\alpha| < \varepsilon, \ \alpha \neq 0 \quad \Rightarrow \quad |o(\alpha^2)| < \frac{1}{2} d^T \nabla^2 f(x^*) d \alpha^2$$

and for these values of α we deduce from (1.17) that $f(x^* + \alpha d) > f(x^*)$.

To conclude that x^* is a (strict) local minimum, one more technical detail is needed. According to the definition of a local minimum (see page 3), we must show that $f(x^*)$ is the lowest value of f in some ball around x^*. But the term $o(\alpha^2)$ and hence the value of ε in the above construction depend on the choice of the direction d. It is clear that this dependence is continuous, since all the other terms in (1.17) are continuous in d.[2] Also, without loss of generality we can restrict d to be of unit length, and then we can take the minimum of ε over all such d. Since the unit sphere in $\mathbb{R}^n$ is compact, the minimum is well defined (thanks to the Weierstrass Theorem which is discussed below). This minimal value of ε provides the radius of the desired ball around x^* in which the lowest value of f is achieved at x^*.

FEASIBLE DIRECTIONS, GLOBAL MINIMA, AND CONVEX PROBLEMS

The key fact that we used in the previous developments was that for every $d \in \mathbb{R}^n$, points of the form $x^* + \alpha d$ for α sufficiently close to 0 belong to D. This is no longer the case if D has a boundary (e.g., D is a closed ball in $\mathbb{R}^n$) and x^* is a point on this boundary. Such situations do not fit into the unconstrained optimization scenario as we defined it at the beginning of Section 1.2.1; however, for simple enough sets D and with some extra care, a similar analysis is possible. Let us call a vector $d \in \mathbb{R}^n$ a *feasible direction* (at x^*) if $x^* + \alpha d \in D$ for small enough $\alpha > 0$ (see Figure 1.1). If not all directions d are feasible, then the condition $\nabla f(x^*) = 0$ is no longer necessary for optimality. We can still define the function (1.4) for every feasible direction d, but the proof of (1.7) is no longer valid because α is now nonnegative. We leave it to the reader to modify that argument and show that if x^* is a local minimum, then $\nabla f(x^*) \cdot d \geq 0$ for every feasible direction d. As for the second-order necessary condition, the inequality (1.14) is still true if $g'(0) = 0$, which together with (1.10) and (1.15) implies that we must have $d^T \nabla^2 f(x^*) d \geq 0$ for all feasible directions satisfying $\nabla f(x^*) \cdot d = 0$.

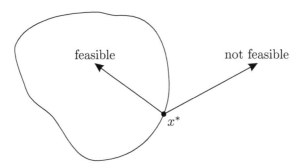

Figure 1.1: Feasible directions

If the set D is *convex*, then the line segment connecting x^* to an arbitrary other point $x \in D$ lies entirely in D. All points on this line segment take the

---

[2]The term $o(\alpha^2)$ can be described more precisely using Taylor's theorem with remainder, which is a higher-order generalization of the Mean Value Theorem; see, e.g., [Rud76, Theorem 5.15]. We will discuss this issue in more detail later when deriving the corresponding result in calculus of variations (see Section 2.6).

form $x^* + \alpha d$, $\alpha \in [0, \bar{\alpha}]$ for some $d \in \mathbb{R}^n$ and $\bar{\alpha} > 0$. This means that the feasible direction approach is particularly suitable for the case of a convex D. But if D is not convex, then the first-order and second-order necessary conditions in terms of feasible directions are conservative. The next exercise touches on the issue of sufficiency.

Exercise 1.1   *Suppose that f is a $\mathcal{C}^2$ function and x^* is a point of its domain at which we have $\nabla f(x^*) \cdot d \geq 0$ and $d^T \nabla^2 f(x^*) d > 0$ for every nonzero feasible direction d. Is x^* necessarily a local minimum of f? Prove or give a counterexample.* $\square$

When we are not dealing with the completely unconstrained case in which D is the entire $\mathbb{R}^n$, we think of D as the constraint set over which f is being minimized. Particularly important in optimization theory is the case when equality constraints are present, so that D is a lower-dimensional surface in $\mathbb{R}^n$ (see Figure 1.2). In such situations, the above method which utilizes feasible directions represented by straight lines is no longer suitable: there might not be any feasible directions, and then the corresponding necessary conditions are vacuous. We will describe a refined approach to constrained optimization in Section 1.2.2; it essentially replaces straight lines with arbitrary curves.

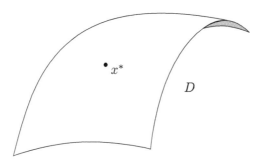

Figure 1.2: A surface

So far we have only discussed local minima. In practice, however, one is typically interested in finding a global minimum over a given domain (or constraint set) D, if such a global minimum exists. We now briefly discuss how conditions for local optimality can be useful for solving global optimization problems as well, provided that these global problems have certain nice features.

The following basic existence result is known as the **Weierstrass Theorem**: *If f is a continuous function and D is a compact set, then there exists a global minimum of f over D.* The reader will recall that for subsets of $\mathbb{R}^n$, compactness can be defined in three equivalent ways:

1) D is compact if it is closed and bounded.

2) D is compact if every open cover of D has a finite subcover.

3) D is compact if every sequence in D has a subsequence converging to some point in D (sequential compactness).

We will revisit compactness and the Weierstrass Theorem in the infinite-dimensional optimization setting.

The necessary conditions for local optimality that we discussed earlier suggest the following procedure for finding a global minimum. First, find all interior points of D satisfying $\nabla f(x^*) = 0$ (the stationary points). If f is not differentiable everywhere, include also points where ∇f does not exist (these points together with the stationary points comprise the *critical points*). Next, find all boundary points satisfying $\nabla f(x^*) \cdot d \geq 0$ for all feasible d. Finally, compare values at all these candidate points and choose the smallest one. If one can afford the computation of second derivatives, then the second-order conditions can be used in combination with the first-order ones.

If D is a convex set and f is a convex function, then the minimization problem is particularly tractable. First, a local minimum is automatically a global one. Second, the first-order necessary condition (for $f \in \mathcal{C}^1$) is also a sufficient condition. Thus if $\nabla f(x^*) \cdot d \geq 0$ for all feasible directions d, or in particular if x^* is an interior point of D and $\nabla f(x^*) = 0$, then x^* is a global minimum. These properties are consequences of the fact (illustrated in Figure 1.3) that the graph of a convex function f lies above that of the linear approximation $x \mapsto f(x^*) + \nabla f(x^*) \cdot (x - x^*)$.

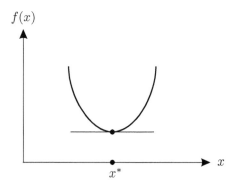

Figure 1.3: A convex function

Efficient numerical algorithms—such as the well-known steepest descent (or gradient) method—exist for converging to points satisfying $\nabla f(x^*) = 0$ (stationary points). For convex problems, these algorithms yield convergence to global minima.

1.2.2 Constrained optimization

Now suppose that D is a surface in $\mathbb{R}^n$ defined by the *equality constraints*

$$h_1(x) = h_2(x) = \cdots = h_m(x) = 0 \qquad (1.18)$$

where h_i, $i = 1, \ldots, m$ are $\mathcal{C}^1$ functions from $\mathbb{R}^n$ to $\mathbb{R}$. We assume that f is a $\mathcal{C}^1$ function and study its minima over D.

FIRST-ORDER NECESSARY CONDITION (LAGRANGE MULTIPLIERS)

Let $x^* \in D$ be a local minimum of f over D. We assume that x^* is a *regular point* of D in the sense that the gradients ∇h_i, $i = 1, \ldots, m$ are linearly independent at x^*. This is a technical assumption needed to rule out degenerate situations; see Exercise 1.2 below.

Instead of line segments containing x^* which we used in the unconstrained case, we now consider *curves* in D passing through x^*. Such a curve is a family of points $x(\alpha) \in D$ parameterized by $\alpha \in \mathbb{R}$, with $x(0) = x^*$. We require the function $x(\cdot)$ to be $\mathcal{C}^1$, at least for α near 0. Given an arbitrary curve of this kind, we can consider the function

$$g(\alpha) := f(x(\alpha)).$$

Note that when there are no equality constraints, functions of the form (1.4) considered previously can be viewed as special cases of this more general construction. From the fact that 0 is a minimum of g, we derive exactly as before that (1.7) holds, i.e., $g'(0) = 0$. To interpret this result in terms of f, note that

$$g'(\alpha) = \nabla f(x(\alpha)) \cdot x'(\alpha)$$

which for $\alpha = 0$ gives

$$g'(0) = \nabla f(x^*) \cdot x'(0) = 0. \qquad (1.19)$$

The vector $x'(0) \in \mathbb{R}^n$ is an important object for us here. From the first-order Taylor expansion $x(\alpha) = x^* + x'(0)\alpha + o(\alpha)$ we see that $x'(0)$ defines a linear approximation of $x(\cdot)$ at x^*. Geometrically, it specifies the infinitesimal direction of the curve (see Figure 1.4). The vector $x'(0)$ is a *tangent vector* to D at x^*. It lives in the *tangent space* to D at x^*, which is denoted by $T_{x^*}D$. (We can think of this space as having its origin at x^*.)

We want to have a more explicit characterization of the tangent space $T_{x^*}D$, which will help us understand it better. Since D was defined as the set of points satisfying the equalities (1.18), and since the points $x(\alpha)$ lie in D by construction, we must have $h_i(x(\alpha)) = 0$ for all α and all $i \in \{1, \ldots, m\}$. Differentiating this formula gives

$$0 = \frac{d}{d\alpha} h_i(x(\alpha)) = \nabla h_i(x(\alpha)) \cdot x'(\alpha), \qquad i = 1, \ldots, m$$

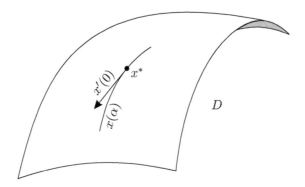

Figure 1.4: A tangent vector

for all α (close enough to 0). Setting $\alpha = 0$ and remembering that $x(0) = x^*$, we obtain

$$0 = \frac{d}{d\alpha}\bigg|_{\alpha=0} h_i(x(\alpha)) = \nabla h_i(x^*) \cdot x'(0), \qquad i = 1, \ldots, m.$$

We have shown that for an arbitrary $\mathcal{C}^1$ curve $x(\cdot)$ in D with $x(0) = x^*$, its tangent vector $x'(0)$ must satisfy $\nabla h_i(x^*) \cdot x'(0) = 0$ for each i. Actually, one can show that the converse is also true, namely, every vector $d \in \mathbb{R}^n$ satisfying

$$\nabla h_i(x^*) \cdot d = 0, \qquad i = 1, \ldots, m \qquad (1.20)$$

is a tangent vector to D at x^* corresponding to some curve. (We do not give a proof of this fact but note that it relies on x^* being a regular point of D.) In other words, the tangent vectors to D at x^* are exactly the vectors d for which (1.20) holds. This is the characterization of the tangent space $T_{x^*}D$ that we were looking for. It is clear from (1.20) that $T_{x^*}D$ is a subspace of $\mathbb{R}^n$; in particular, if d is a tangent vector, then so is $-d$ (going from $x'(0)$ to $-x'(0)$ corresponds to reversing the direction of the curve).

Now let us go back to (1.19), which tells us that $\nabla f(x^*) \cdot d = 0$ for all $d \in T_{x^*}D$ (since the curve $x(\cdot)$ and thus the tangent vector $x'(0)$ were arbitrary). In view of the characterization of $T_{x^*}D$ given by (1.20), we can rewrite this condition as follows:

$$\nabla f(x^*) \cdot d = 0 \qquad \forall d \text{ such that } \nabla h_i(x^*) \cdot d = 0, \ i = 1, \ldots, m. \qquad (1.21)$$

The relation between $\nabla f(x^*)$ and $\nabla h_i(x^*)$ expressed by (1.21) looks somewhat clumsy, since checking it involves a search over d. Can we eliminate d from this relation and make it more explicit? A careful look at (1.21) should quickly lead the reader to the following statement.

Claim: The gradient of f at x^* is a linear combination of the gradients of the constraint functions $h_1, \ldots, h_m$ at x^*:

$$\nabla f(x^*) \in \text{span}\{\nabla h_i(x^*), \ i = 1, \ldots, m\}. \qquad (1.22)$$

Indeed, if the claim were not true, then $\nabla f(x^*)$ would have a component orthogonal to span$\{\nabla h_i(x^*)\}$, i.e, there would exist a $d \neq 0$ satisfying (1.20) such that $\nabla f(x^*)$ can be written in the form

$$\nabla f(x^*) = d - \sum_{i=1}^{m} \lambda_i^* \nabla h_i(x^*) \tag{1.23}$$

for some $\lambda_1^*, \ldots, \lambda_m^* \in \mathbb{R}$. Taking the inner product with d on both sides of (1.23) and using (1.20) gives

$$\nabla f(x^*) \cdot d = d \cdot d \neq 0$$

and we reach a contradiction with (1.21).

Geometrically, the claim says that $\nabla f(x^*)$ is normal to D at x^*. This situation is illustrated in Figure 1.5 for the case of two constraints in $\mathbb{R}^3$. Note that if there is only one constraint, say $h_1(x) = 0$, then D is a two-dimensional surface and $\nabla f(x^*)$ must be proportional to $\nabla h_1(x^*)$, the normal direction to D at x^*. When the second constraint $h_2(x) = 0$ is added, D becomes a curve (the thick curve in the figure) and $\nabla f(x^*)$ is allowed to live in the plane spanned by $\nabla h_1(x^*)$ and $\nabla h_2(x^*)$, i.e., the normal plane to D at x^*. In general, the intuition behind the claim is that unless $\nabla f(x^*)$ is normal to D, there are curves in D passing through x^* whose tangent vectors at x^* make both positive and negative inner products with $\nabla f(x^*)$, hence in particular f can be decreased by moving away from x^* while staying in D.

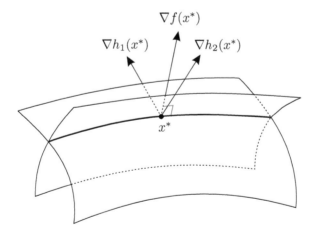

Figure 1.5: Gradient vectors and constrained optimality

The condition (1.22) means that there exist real numbers $\lambda_1^*, \ldots, \lambda_m^*$ such that

$$\boxed{\nabla f(x^*) + \lambda_1^* \nabla h_1(x^*) + \cdots + \lambda_m^* \nabla h_m(x^*) = 0} \tag{1.24}$$

This is the **first-order necessary condition for constrained optimality**. The coefficients λ_i^*, $i = 1, \ldots, m$ are called *Lagrange multipliers*.

Exercise 1.2   *Give an example where a local minimum x^* is not a regular point and the above necessary condition is false (be sure to justify both of these claims).*   □

The above proof of the first-order necessary condition for constrained optimality involves geometric concepts. We also left a gap in it because we did not prove the converse implication in the equivalent characterization of the tangent space given by (1.20). We now give a shorter alternative proof which is purely analytic, and which will be useful when we study problems with constraints in calculus of variations. However, the geometric intuition behind the previous proof will be helpful for us later as well. We invite the reader to study both proofs as a way of testing the mathematical background knowledge that will be required in the subsequent chapters.

Let us start again by assuming that x^* is a local minimum of f over D, where D is a surface in $\mathbb{R}^n$ defined by the equality constraints (1.18) and x^* is a regular point of D. Our goal is to rederive the necessary condition expressed by (1.24). For simplicity, we only give the argument for the case of a single constraint $h(x) = 0$, i.e., $m = 1$; the extension to $m > 1$ is straightforward (see Exercise 1.3 below). Given two arbitrary vectors $d_1, d_2 \in \mathbb{R}^n$, we can consider the following map from $\mathbb{R} \times \mathbb{R}$ to itself:

$$F : (\alpha_1, \alpha_2) \mapsto \big(f(x^* + \alpha_1 d_1 + \alpha_2 d_2), h(x^* + \alpha_1 d_1 + \alpha_2 d_2)\big).$$

The Jacobian matrix of F at $(0, 0)$ is

$$\begin{pmatrix} \nabla f(x^*) \cdot d_1 & \nabla f(x^*) \cdot d_2 \\ \nabla h(x^*) \cdot d_1 & \nabla h(x^*) \cdot d_2 \end{pmatrix}. \tag{1.25}$$

If this Jacobian matrix were nonsingular, then we could apply the Inverse Function Theorem (see, e.g., [Rud76, Theorem 9.24]) and conclude that there are neighborhoods of $(0, 0)$ and $F(0, 0) = (f(x^*), 0)$ on which the map F is a bijection (has an inverse). This would imply, in particular, that there are points x arbitrarily close to x^* such that $h(x) = 0$ and $f(x) < f(x^*)$; such points would be obtained by taking preimages of points on the ray directed to the left from $F(0, 0)$ in Figure 1.6. But this cannot be true, since $h(x) = 0$ means that $x \in D$ and we know that x^* is a local minimum of f over D. Therefore, the matrix (1.25) is singular.

Regularity of x^* in the present case just means that the gradient $\nabla h(x^*)$ is nonzero. Choose a d_1 such that $\nabla h(x^*) \cdot d_1 \neq 0$. With this d_1 fixed, let $\lambda^* := -(\nabla f(x^*) \cdot d_1)/(\nabla h(x^*) \cdot d_1)$, so that $\nabla f(x^*) \cdot d_1 = -\lambda^* \nabla h(x^*) \cdot d_1$. Since the matrix (1.25) must be singular for all choices of d_2, its first row must be a constant multiple of its second row (the second row being nonzero

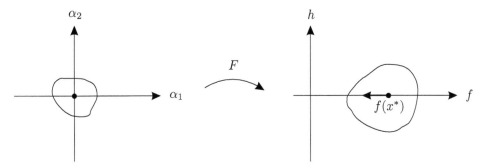

Figure 1.6: Illustrating the alternative proof

by our choice of d_1). Thus we have $\nabla f(x^*) \cdot d_2 = -\lambda^* \nabla h(x^*) \cdot d_2$, or $(\nabla f(x^*) + \lambda^* \nabla h(x^*)) \cdot d_2 = 0$, and this must be true for all $d_2 \in \mathbb{R}^n$. It follows that $\nabla f(x^*) + \lambda^* \nabla h(x^*) = 0$, which proves (1.24) for the case when $m = 1$.

Exercise 1.3   *Generalize the previous argument to an arbitrary number $m \geq 1$ of equality constraints (still assuming that x^* is a regular point).*   □

The first-order necessary condition for constrained optimality generalizes the corresponding result we derived earlier for the unconstrained case. The condition (1.24) together with the constraints (1.18) is a system of $n + m$ equations in $n + m$ unknowns: n components of x^* plus m components of the Lagrange multiplier vector $\lambda^* = (\lambda_1^*, \dots, \lambda_m^*)^T$. For $m = 0$, we recover the condition (1.11) which consists of n equations in n unknowns. To make this relation even more explicit, consider the function $\ell : \mathbb{R}^n \times \mathbb{R}^m \to \mathbb{R}$ defined by

$$\ell(x, \lambda) := f(x) + \sum_{i=1}^{m} \lambda_i h_i(x) \qquad (1.26)$$

which we call the *augmented cost* function. If x^* is a local constrained minimum of f and λ^* is the corresponding vector of Lagrange multipliers for which (1.24) holds, then the gradient of ℓ at (x^*, λ^*) satisfies

$$\nabla \ell(x^*, \lambda^*) = \begin{pmatrix} \ell_x(x^*, \lambda^*) \\ \ell_\lambda(x^*, \lambda^*) \end{pmatrix} = \begin{pmatrix} \nabla f(x^*) + \sum_{i=1}^{m} \lambda_i^* \nabla h_i(x^*) \\ h(x^*) \end{pmatrix} = 0 \quad (1.27)$$

where ℓ_x, ℓ_λ are the vectors of partial derivatives of ℓ with respect to the components of x and λ, respectively, and $h = (h_1, \dots, h_m)^T$ is the vector of constraint functions. We conclude that (x^*, λ^*) is a usual (unconstrained) stationary point of the augmented cost ℓ. Loosely speaking, adding Lagrange multipliers converts a constrained problem into an unconstrained one, and the first-order necessary condition (1.24) for constrained optimality is recovered from the first-order necessary condition for unconstrained optimality applied to ℓ.

The idea of passing from constrained minimization of the original cost function to unconstrained minimization of the augmented cost function is due to Lagrange. If (x^*, λ^*) is a minimum of ℓ, then we must have $h(x^*) = 0$ (because otherwise we could decrease ℓ by changing λ^*), and subject to these constraints x^* should minimize f (because otherwise it would not minimize ℓ). Also, it is clear that (1.27) must hold. However, it does *not* follow that (1.27) is a necessary condition for x^* to be a constrained minimum of f. Unfortunately, there is no quick way to derive the first-order necessary condition for constrained optimality by working with the augmented cost—something that Lagrange originally attempted to do. Nevertheless, the basic form of the augmented cost function (1.26) is fundamental in constrained optimization theory, and will reappear in various forms several times in this book.

Even though the condition in terms of Lagrange multipliers is only necessary and not sufficient for constrained optimality, it is very useful for narrowing down candidates for local extrema. The next exercise illustrates this point for a well-known optimization problem arising in optics.

Exercise 1.4   *Consider a curve D in the plane described by the equation $h(x) = 0$, where $h : \mathbb{R}^2 \to \mathbb{R}$ is a C^1 function. Let y and z be two fixed points in the plane, lying on the same side with respect to D (but not on D itself). Suppose that a ray of light emanates from y, gets reflected off D at some point $x^* \in D$, and arrives at z. Consider the following two statements: (i) x^* must be such that the total Euclidean distance traveled by light to go from y to z is minimized over all nearby candidate reflection points $x \in D$ (Fermat's principle); (ii) the angles that the light ray makes with the line normal to D at x^* before and after the reflection must be the same (the law of reflection). Accepting the first statement as a hypothesis, prove that the second statement follows from it, with the help of the first-order necessary condition for constrained optimality (1.24).*   □

SECOND-ORDER CONDITIONS

For the sake of completeness, we quickly state the second-order conditions for constrained optimality; they will not be used in the sequel. For the necessary condition, suppose that x^* is a regular point of D and a local minimum of f over D, where D is defined by the equality constraints (1.18) as before. We let λ^* be the vector of Lagrange multipliers provided by the first-order necessary condition, and define the augmented cost ℓ as in (1.26). We also assume that f is C^2. Consider the Hessian of ℓ with respect to x evaluated at (x^*, λ^*):

$$\ell_{xx}(x^*, \lambda^*) = \nabla^2 f(x^*) + \sum_{i=1}^{m} \lambda_i^* \nabla^2 h_i(x^*).$$

The second-order necessary condition says that this Hessian matrix must be positive semidefinite on the tangent space to D at x^*, i.e., we must have $d^T \ell_{xx}(x^*, \lambda^*)d \geq 0$ for all $d \in T_{x^*}D$. Note that this is weaker than asking the above Hessian matrix to be positive semidefinite in the usual sense (on the entire $\mathbb{R}^n$).

The second-order sufficient condition says that a point $x^* \in D$ is a strict constrained local minimum of f if the first-order necessary condition for constrained optimality (1.24) holds and, in addition, we have

$$d^T \ell_{xx}(x^*, \lambda^*)d > 0 \qquad \forall d \text{ such that } \nabla h_i(x^*) \cdot d = 0, \ i = 1, \ldots, m. \quad (1.28)$$

Again, here λ^* is the vector of Lagrange multipliers and ℓ is the corresponding augmented cost. Note that regularity of x^* is not needed for this sufficient condition to be true. If x^* is in fact a regular point, then we know (from our derivation of the first-order necessary condition for constrained optimality) that the condition imposed on d in (1.28) describes exactly the tangent vectors to D at x^*. In other words, in this case (1.28) is equivalent to saying that $\ell_{xx}(x^*, \lambda^*)$ is positive definite on the tangent space $T_{x^*}D$.

1.3   PREVIEW OF INFINITE-DIMENSIONAL OPTIMIZATION

In Section 1.2 we considered the problem of minimizing a function $f : \mathbb{R}^n \to \mathbb{R}$. Now, instead of $\mathbb{R}^n$ we want to allow a general vector space V, and in fact we are interested in the case when this vector space V is infinite-dimensional. Specifically, V will itself be a space of functions. Let us denote a generic function in V by y, reserving the letter x for the argument of y. (This x will typically be a scalar, and has no relation with $x \in \mathbb{R}^n$ from the previous section.) The function to be minimized is a real-valued function on V, which we now denote by J. Since J is a function on a space of functions, it is called a *functional*. To summarize, we are minimizing a functional $J : V \to \mathbb{R}$.

Unlike in the case of $\mathbb{R}^n$, there does not exist a "universal" function space. Many different choices for V are possible, and specifying the desired space V is part of the problem formulation. Another issue is that in order to define *local* minima of J over V, we need to specify what it means for two functions in V to be close to each other. Recall that in the definition of a local minimum in Section 1.2, a ball of radius ε with respect to the standard Euclidean norm on $\mathbb{R}^n$ was used to define the notion of closeness. In the present case we will again employ ε-balls, but we need to specify which norm we are going to use. While in $\mathbb{R}^n$ all norms are equivalent (i.e., are within a constant multiple of one another), in function spaces different choices of a norm lead to drastically different notions of closeness. Thus, the first thing we need to do is become more familiar with function spaces and norms on them.

1.3.1   Function spaces, norms, and local minima

Typical function spaces that we will consider are spaces of functions from some interval $[a, b]$ to $\mathbb{R}^n$ (for some $n \geq 1$). Different spaces result from placing different requirements on the regularity of these functions. For example, we will frequently work with the function space $\mathcal{C}^k([a, b], \mathbb{R}^n)$, whose elements are k-times continuously differentiable (here $k \geq 0$ is an integer; for $k = 0$ the functions are just continuous). Relaxing the $\mathcal{C}^k$ assumption, we can arrive at the spaces of piecewise continuous functions or even measurable functions (we will define these more precisely later when we need them). On the other hand, stronger regularity assumptions lead us to $\mathcal{C}^\infty$ (smooth, or infinitely many times differentiable) functions or to real analytic functions (the latter are $\mathcal{C}^\infty$ functions that agree with their Taylor series around every point).

We regard these function spaces as linear vector spaces over $\mathbb{R}$. Why are they infinite-dimensional? One way to see this is to observe that the monomials $1, x, x^2, x^3, \ldots$ are linearly independent. Another example of an infinite set of linearly independent functions is provided by the (trigonometric) Fourier basis.

As we already mentioned, we also need to equip our function space V with a *norm* $\| \cdot \|$. This is a real-valued function on V which is positive definite ($\|y\| > 0$ if $y \not\equiv 0$), homogeneous ($\|\lambda y\| = |\lambda| \cdot \|y\|$ for all $\lambda \in \mathbb{R}$, $y \in V$), and satisfies the triangle inequality ($\|y + z\| \leq \|y\| + \|z\|$). The norm gives us the notion of a *distance*, or *metric*, $d(y, z) := \|y - z\|$. This allows us to define local minima and enables us to talk about topological concepts such as convergence and continuity (more on this in Section 1.3.4 below). We will see how the norm plays a crucial role in the subsequent developments.

On the space $\mathcal{C}^0([a, b], \mathbb{R}^n)$, a commonly used norm is

$$\|y\|_0 := \max_{a \leq x \leq b} |y(x)| \tag{1.29}$$

where $| \cdot |$ is the standard Euclidean norm on $\mathbb{R}^n$ as before. Replacing the maximum by a supremum, we can extend the 0-norm (1.29) to functions that are defined over an infinite interval or are not necessarily continuous. On $\mathcal{C}^1([a, b], \mathbb{R}^n)$, another natural candidate for a norm is obtained by adding the 0-norms of y and its first derivative:

$$\|y\|_1 := \max_{a \leq x \leq b} |y(x)| + \max_{a \leq x \leq b} |y'(x)|. \tag{1.30}$$

This construction can be continued in the obvious way to yield the k-norm on $\mathcal{C}^k([a, b], \mathbb{R}^n)$ for each k. The k-norm can also be used on $\mathcal{C}^\ell([a, b], \mathbb{R}^n)$ for all $\ell \geq k$. There exist many other norms, such as for example the $\mathcal{L}_p$

norm

$$\|y\|_{\mathcal{L}_p} := \left(\int_a^b |y(x)|^p dx \right)^{1/p} \tag{1.31}$$

where p is a positive integer. In fact, the 0-norm (1.29) is also known as the $\mathcal{L}_\infty$ norm.

We are now ready to formally define local minima of a functional. Let V be a vector space of functions equipped with a norm $\|\cdot\|$, let A be a subset of V, and let J be a real-valued functional defined on V (or just on A). A function $y^* \in A$ is a *local minimum* of J over A if there exists an $\varepsilon > 0$ such that for all $y \in A$ satisfying $\|y - y^*\| < \varepsilon$ we have

$$J(y^*) \leq J(y).$$

Note that this definition of a local minimum is completely analogous to the one in the previous section, modulo the change of notation $x \mapsto y$, $D \mapsto A$, $f \mapsto J$, $|\cdot| \mapsto \|\cdot\|$ (also, implicitly, $\mathbb{R}^n \mapsto V$). Strict minima, global minima, and the corresponding notions of maxima are defined in the same way as before. We will continue to refer to minima and maxima collectively as *extrema*.

For the norm $\|\cdot\|$, we will typically use either the 0-norm (1.29) or the 1-norm (1.30), with V being $\mathcal{C}^0([a, b], \mathbb{R}^n)$ or $\mathcal{C}^1([a, b], \mathbb{R}^n)$, respectively. In the remainder of this section we discuss some general conditions for optimality which apply to both of these norms. However, when we develop more specific results later in calculus of variations, our findings for these two cases will be quite different.

1.3.2 First variation and first-order necessary condition

To develop the first-order necessary condition for optimality, we need a notion of derivative for functionals. Let $J : V \to \mathbb{R}$ be a functional on a function space V, and consider some function $y \in V$. The derivative of J at y, which will now be called the first variation, will also be a functional on V, and in fact this functional will be linear. To define it, we consider functions in V of the form $y + \alpha\eta$, where $\eta \in V$ and α is a real parameter (which can be restricted to some interval around 0). The reader will recognize these functions as infinite-dimensional analogs of the points $x^* + \alpha d$ around a given point $x^* \in \mathbb{R}^n$, which we utilized earlier.

A linear functional $\delta J|_y : V \to \mathbb{R}$ is called the *first variation* of J at y if for all η and all α we have

$$J(y + \alpha\eta) = J(y) + \delta J|_y (\eta)\alpha + o(\alpha) \tag{1.32}$$

where $o(\alpha)$ satisfies (1.6). The somewhat cumbersome notation $\delta J|_y (\eta)$ is meant to emphasize that the linear term in α in the expansion (1.32) depends on both y and η. The requirement that $\delta J|_y$ must be a linear functional is understood in the usual sense: $\delta J|_y (\alpha_1 \eta_1 + \alpha_2 \eta_2) = \alpha_1 \, \delta J|_y (\eta_1) + \alpha_2 \, \delta J|_y (\eta_2)$ for all $\eta_1, \eta_2 \in V$ and $\alpha_1, \alpha_2 \in \mathbb{R}$.

The first variation as defined above corresponds to the so-called Gateaux derivative of J, which is just the usual derivative of $J(y + \alpha \eta)$ with respect to α (for fixed y and η) evaluated at $\alpha = 0$:

$$\delta J|_y (\eta) = \lim_{\alpha \to 0} \frac{J(y + \alpha \eta) - J(y)}{\alpha}. \tag{1.33}$$

In other words, if we define

$$g(\alpha) := J(y + \alpha \eta) \tag{1.34}$$

then

$$\delta J|_y (\eta) = g'(0) \tag{1.35}$$

and (1.32) reduces exactly to our earlier first-order expansion (1.5).

Now, suppose that y^* is a local minimum of J over some subset A of V. We call a perturbation[3] $\eta \in V$ *admissible* (with respect to the subset A) if $y^* + \alpha \eta \in A$ for all α sufficiently close to 0. It follows from our definitions of a local minimum and an admissible perturbation that $J(y^* + \alpha \eta)$ as a function of α has a local minimum at $\alpha = 0$ for each admissible η. Let us assume that the first variation $\delta J|_{y*}$ exists (which is of course not always the case) so that we have (1.32). Applying the same reasoning that we used to derive the necessary condition (1.7) on the basis of (1.5), we quickly arrive at the **first-order necessary condition for optimality**: *For all admissible perturbations η, we must have*

$$\boxed{\delta J|_{y^*} (\eta) = 0} \tag{1.36}$$

As in the finite-dimensional case, the first-order necessary condition applies to both minima and maxima.

When we were studying a minimum x^* of $f : \mathbb{R}^n \to \mathbb{R}$ with the help of the function $g(\alpha) := f(x^* + \alpha d)$, it was easy to translate the equality $g'(0) = 0$ via the formula (1.10) into the necessary condition $\nabla f(x^*) = 0$. The necessary condition (1.36), while conceptually very similar, is much less constructive. To be able to apply it, we need to learn how to compute the first variation of some useful functionals. This subject will be further discussed in the next chapter; for now, we offer an example for the reader to work out.

---

[3]With a slight abuse of terminology, we call η a perturbation even though the actual perturbation is $\alpha \eta$.

Exercise 1.5   *Consider the space $V = \mathcal{C}^0([0,1], \mathbb{R})$, let $\varphi : \mathbb{R} \to \mathbb{R}$ be a $\mathcal{C}^1$ function, and define the functional J on V by $J(y) = \int_0^1 \varphi(y(x))dx$. Show that its first variation exists and is given by the formula $\delta J|_y(\eta) = \int_0^1 \varphi'(y(x))\eta(x)dx$.* $\square$

Our notion of the first variation, defined via the expansion (1.32), is independent of the choice of the norm on V. This means that the first-order necessary condition (1.36) is valid for every norm. To obtain a necessary condition better tailored to a particular norm, we could define $\delta J|_y$ differently, by using the following expansion instead of (1.32):

$$J(y + \eta) = J(y) + \delta J|_y(\eta) + o(\|\eta\|). \tag{1.37}$$

The difference with our original formulation is subtle but substantial. The earlier expansion (1.32) describes how the value of J changes with α for each fixed η. In (1.37), the higher-order term is a function of $\|\eta\|$ and so the expansion captures the effect of all η at once (while α is no longer needed). We remark that the first variation defined via (1.37) corresponds to the so-called Fréchet derivative of J, which is a stronger differentiability notion than the Gateaux derivative (1.33). In fact, (1.37) suggests constructing more general perturbations: instead of working with functions of the form $y + \alpha\eta$, where η is fixed and α is a scalar parameter, we can consider perturbed functions $y + \eta$ which can approach y in a more arbitrary manner as $\|\eta\|$ tends to 0 (multiplying η by a vanishing parameter is just one possibility). This generalization is conceptually similar to that of passing from the lines $x^* + \alpha d$ used in Section 1.2.1 to the curves $x(\alpha)$ utilized in Section 1.2.2. We will start seeing perturbations of this kind in Chapter 3.

In what follows, we retain our original definition of the first variation in terms of (1.32). It is somewhat simpler to work with and is adequate for our needs (at least through Chapter 2). While the norm-dependent formulation could potentially provide sharper conditions for optimality, it takes more work to verify (1.37) for all η compared to verifying (1.32) for a fixed η. Besides, we will eventually abandon the analysis based on the first variation altogether in favor of more powerful tools. However, it is useful to be aware of the alternative formulation (1.37), and we will occasionally make some side remarks related to it. This issue will resurface in Chapter 3 where, although the alternative definition (1.37) of the first variation will not be specifically needed, we will use more general perturbations along the lines of the preceding discussion.

1.3.3   Second variation and second-order conditions

A real-valued functional B on $V \times V$ is called *bilinear* if it is linear in each argument (when the other one is fixed). Setting $Q(y) := B(y, y)$ we then obtain a *quadratic functional*, or *quadratic form*, on V. This is a direct generalization of the corresponding familiar concepts for finite-dimensional vector spaces.

A quadratic form $\delta^2 J\big|_y : V \to \mathbb{R}$ is called the *second variation* of J at y if for all $\eta \in V$ and all α we have

$$J(y + \alpha\eta) = J(y) + \delta J\big|_y (\eta)\alpha + \delta^2 J\big|_y (\eta)\alpha^2 + o(\alpha^2). \qquad (1.38)$$

This exactly corresponds to our previous second-order expansion (1.12) for the function g given by (1.34). Repeating the same argument we used earlier to prove (1.14), we easily establish the following **second-order necessary condition for optimality**: *If y^* is a local minimum of J over $A \subset V$, then for all admissible perturbations η we have*

$$\boxed{\delta^2 J\big|_{y^*} (\eta) \geq 0} \qquad (1.39)$$

In other words, the second variation of J at y^* must be positive semidefinite on the space of admissible perturbations. For local maxima, the inequality in (1.39) is reversed. Of course, the usefulness of the condition will depend on our ability to compute the second variation of the functionals that we will want to study.

Exercise 1.6 *Consider the same functional J as in Exercise 1.5, but assume now that φ is C^2. Derive a formula for the second variation of J (make sure that it is indeed a quadratic form).* □

What about a second-order *sufficient* condition for optimality? By analogy with the second-order sufficient condition (1.16) which we derived for the finite-dimensional case, we may guess that we need to combine the first-order necessary condition (1.36) with the strict-inequality counterpart of the second-order necessary condition (1.39), i.e.,

$$\delta^2 J\big|_{y^*} (\eta) > 0 \qquad (1.40)$$

(this should again hold for all admissible perturbations η with respect to a subset A of V over which we want y^* to be a minimum). We would then hope to show that for $y = y^*$ the second-order term in (1.38) dominates the higher-order term $o(\alpha^2)$, which would imply that y^* is a strict local minimum (since the first-order term is 0). Our earlier proof of sufficiency of (1.16) followed the same idea. However, examining that proof more closely, the reader will discover that in the present case the argument does not go through.

We know that there exists an $\varepsilon > 0$ such that for all nonzero α with $|\alpha| < \varepsilon$ we have $|o(\alpha^2)| < \delta^2 J\big|_{y^*} (\eta)\alpha^2$. Using this inequality and (1.36), we obtain from (1.38) that $J(y^* + \alpha\eta) > J(y^*)$. Note that this does not yet prove that y^* is a (strict) local minimum of J. According to the definition of a local minimum, we must show that $J(y^*)$ is the lowest value of J in some ball around y^* with respect to the selected norm $\|\cdot\|$ on V. The problem is that the term $o(\alpha^2)$ and hence the above ε depend on the choice of the perturbation η. In the finite-dimensional case we took the minimum of ε over all perturbations of unit length, but we cannot do that here because the unit sphere in the infinite-dimensional space V is not compact and the Weierstrass Theorem does not apply to it (see Section 1.3.4 below).

One way to resolve the above difficulty would be as follows. The first step is to strengthen the condition (1.40) to

$$\delta^2 J\big|_{y^*}(\eta) \geq \lambda \|\eta\|^2 \tag{1.41}$$

for some number $\lambda > 0$. The property (1.41) does not automatically follow from (1.40), again because we are in an infinite-dimensional space. (Quadratic forms satisfying (1.41) are sometimes called uniformly positive definite.) The second step is to modify the definitions of the first and second variations by explicitly requiring that the higher-order terms decay uniformly with respect to $\|\eta\|$. We already mentioned such an alternative definition of the first variation via the expansion (1.37). Similarly, we could define $\delta^2 J\big|_y$ via the following expansion in place of (1.38):

$$J(y + \eta) = J(y) + \delta J\big|_y(\eta) + \delta^2 J\big|_y(\eta) + o(\|\eta\|^2). \tag{1.42}$$

Adopting these alternative definitions and assuming that (1.36) and (1.41) hold, we could easily prove optimality by noting that $|o(\|\eta\|^2)| < \lambda \|\eta\|^2$ when $\|\eta\|$ is small enough.

With our current definitions of the first and second variations in terms of (1.32) and (1.38), we do not have a general second-order sufficient condition for optimality. However, in variational problems that we are going to study, the functional J to be minimized will take a specific form. This additional structure will allow us to derive conditions under which second-order terms dominate higher-order terms, resulting in optimality. The above discussion was given mainly for illustrative purposes, and will not be directly used in the sequel.

1.3.4   Global minima and convex problems

Regarding global minima of J over a set $A \subset V$, much of the discussion on global minima given at the end of Section 1.2.1 carries over to the present case. In particular, the Weierstrass Theorem is still valid, provided that compactness of A is understood in the sense of the second or third definition given on page 10 (existence of finite subcovers or sequential compactness). These two definitions of compactness are equivalent for linear vector spaces equipped with a norm (or, more generally, a metric). On the other hand, closed and bounded subsets of an infinite-dimensional vector space are not necessarily compact—we already mentioned noncompactness of the unit sphere—and the Weierstrass Theorem does not apply to them; see the next exercise. We note that since our function space V has a norm, the notions of continuity of J and convergence, closedness, boundedness, and openness in V with respect to this norm are defined exactly as their familiar counterparts in $\mathbb{R}^n$. We leave it to the reader to write down precise definitions or consult the references given at the end of this chapter.

Exercise 1.7   *Give an example of a function space V, a norm on V, a closed and bounded subset A of V, and a continuous functional J on V such that a global minimum of J over A does not exist. (Be sure to demonstrate that all the requested properties hold.)*   □

If J is a convex functional and $A \subset V$ is a convex set, then the optimization problem enjoys the same properties as the ones mentioned at the end of Section 1.2.1 for finite-dimensional convex problems. Namely, a local minimum is automatically a global one, and the first-order necessary condition is also a sufficient condition for a minimum. (Convexity of a functional and convexity of a subset of an infinite-dimensional linear vector space are defined exactly as the corresponding standard notions in the finite-dimensional case.) However, imposing extra assumptions to ensure convexity of J would severely restrict the classes of problems that we want to study. In this book, we focus on general theory that applies to not necessarily convex problems; we will not directly use results from convex optimization. Nevertheless, some basic concepts from (finite-dimensional) convex analysis will be important for us later, particularly when we derive the maximum principle.

1.4   NOTES AND REFERENCES FOR CHAPTER 1

Success stories of optimal control theory in various applications are too numerous to be listed here; see [CEHS87, Cla10, ST05, Swa84] for some examples from engineering and well beyond. The reader interested in applications will easily find many other references.

The material in Section 1.2 can be found in standard texts on optimization, such as [Lue84] or [Ber99]. See also Sections 5.2-5.4 of the book [AF66], which will be one of our main references for the optimal control chapters. Complete proofs of the results presented in Section 1.2.2, including the fact that the condition (1.20) is sufficient for d to be a tangent vector, are given in [Lue84, Chapter 10]. The alternative argument based on the inverse function theorem is adopted from [Mac05, Section 1.4]. The necessary condition in terms of Lagrange multipliers can also be derived from a cone separation property (via Farkas's lemma) as shown, e.g., in [Ber99, Section 3.3.6]; we will see this type of reasoning in the proof of the maximum principle in Chapter 4.

Section 1.3 is largely based on [GF63], which will be our main reference for calculus of variations; function spaces, functionals, and the first variation are introduced in the first several sections of that book, while the second variation is discussed later in Chapter 5. Essentially the same material but in condensed form can be found in [AF66, Section 5.5]. In [GF63] as well as in [Mac05] the first and second variations are defined via (1.37) and (1.42), while other sources such as [You80] follow the approach based on (1.32) and (1.38).

For further background on function spaces and relevant topological concepts, the reader can consult [Rud76] or [Sut75] (the latter text is somewhat more advanced). Another recommended reference on these topics is [Lue69], where Gateaux and Fréchet derivatives and their role in functional minimization are also thoroughly discussed. A general treatment of convex functionals and their minimization is given in [Lue69, Chapter 7]; for convexity-based results more specific to calculus of variations and optimal control, see the monograph [Roc74], the more recent papers [RW00] and [GS07], and the references therein.

Chapter Two

Calculus of Variations

2.1 EXAMPLES OF VARIATIONAL PROBLEMS

We begin this chapter by describing a few examples of classical variational problems, i.e., problems in which a path is to be chosen from a given family of admissible paths so as to minimize the value of some functional. These examples will help us motivate the general problem formulation and will serve as benchmarks for the theory that we will be developing.

2.1.1 Dido's isoperimetric problem

According to a legend about the foundation of Carthage (around 850 B.C.), Dido purchased from a local king the land along the North African coastline that could be enclosed by the hide of an ox. She sliced the hide into very thin strips, tied them together, and was able to enclose a sizable area which became the city of Carthage.

Let us formulate this as an optimization problem. Assume that the coast is a straight line (represented by the x-axis in Figure 2.1). The hide strips tied together correspond to a curve of a fixed length. The problem is to maximize the area under this curve.

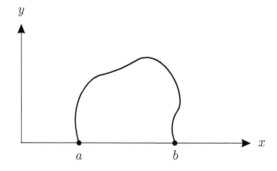

Figure 2.1: Dido's problem

More formally, admissible curves are graphs of continuous functions[1] $y : [a, b] \rightarrow \mathbb{R}$ satisfying the endpoint constraints $y(a) = y(b) = 0$. The area under such a curve is

$$J(y) = \int_a^b y(x)dx$$

which is the quantity to be maximized. Of course, it is easy to convert this to a minimization problem by working with $-J$. Keeping in mind the original problem context, we should actually restrict y to be nonnegative (so it does not go into the ocean) or at least replace y by $\max\{0, y\}$ in the definition of J; but even without these modifications, it is clear that only nonnegative y can be optimal. Assuming that the curve is differentiable (at least almost everywhere), we can write the length constraint as

$$\int_a^b \sqrt{1 + (y'(x))^2}dx = C_0 \tag{2.1}$$

where C_0 is a fixed constant. Indeed, by the Pythagorean Theorem the integrand in (2.1) is the arclength element.

At this point the reader can probably guess what the optimal curve should be. It is an arc of a circle. This solution was known to Zenodorus (2nd century B.C.), although its rigorous derivation requires tools from calculus of variations which will be presented in this chapter.

Note that since we are representing admissible curves by graphs of functions y, we are excluding circular arcs that are longer than a semicircle. In the context of Dido's problem, we can think of the interval $[a, b]$ as not actually being specified in advance but instead as being chosen based on the length constraint. Then, once we know that the optimal curve must be a circular arc, it is straightforward to check that the optimal choice of the interval is the one in which the circular arc of length C_0 is precisely a semicircle, with the interval serving as the circle's diameter (and thus having length $2C_0/\pi$).

2.1.2 Light reflection and refraction

In free space, light travels along a path of shortest distance—which is of course a straight line. This is already a solution to a variational problem, albeit a very simple one. More interesting situations arise when a light ray encounters the edge of a medium. Two basic phenomena are *reflection* and *refraction* of light (see Figure 2.2).

---

[1]Actually, the curve in Figure 2.1 is not the graph of a function. There is a tendency to ignore this difference between curves and graphs of functions when formulating calculus of variations problems.

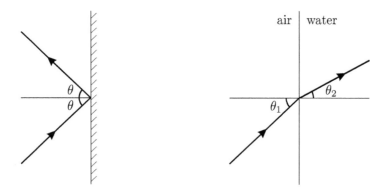

Figure 2.2: Light reflection and refraction

In the case of reflection, Hero of Alexandria (who probably lived in the 1st century A.D.) suggested that light still takes the path of shortest distance among nearby paths. When the reflecting surface is a plane, one can argue using some simple geometry that the angles between the normal to this plane and the light ray before and after the reflection must then be the same. This result generalizes to curved reflecting surfaces, although proving it rigorously is not trivial (see Exercise 1.4 in the previous chapter).

Analyzing refraction is more challenging. Ptolemy made a list of angle pairs (θ_1, θ_2) corresponding to situations like the one depicted on the right in Figure 2.2. His list dates back to 140 A.D. and contains quite a few values. (This is ancient Greek experimental physics!) A pattern in Ptolemy's results was found only much later, in 1621, when Snell stated his law:

$$\sin \theta_1 = n \sin \theta_2 \qquad (2.2)$$

where n is the ratio of the speeds of light in the two media ($n \approx 1.33$ when light passes from air to water). A satisfactory explanation of this behavior was first given by Fermat around 1650. Fermat's principle states that, although the light does not take the path of shortest distance any more, it travels along the path of *shortest time*. Snell's law can be derived from Fermat's principle by differential calculus; in fact, this was one of the examples that Leibniz gave to illustrate the power of calculus in his original 1684 calculus monograph.

The problems of light reflection and refraction are mentioned here mainly for historical reasons, and we do not proceed to mathematically formalize them. (However, we will revisit light reflection in Section 7.4.3.)

2.1.3   Catenary

Suppose that a chain of a given length, with uniform mass density, is suspended freely between two fixed points (see Figure 2.3). What will be the shape of this chain? This question was posed by Galileo in the 1630s, and he claimed—incorrectly—that the solution is a parabola.

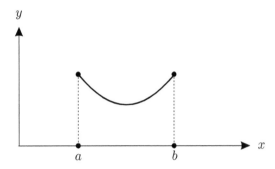

Figure 2.3: A catenary

Mathematically, the chain is described by a continuous function y : $[a, b] \rightarrow [0, \infty)$, and the two suspension points are specified by endpoint constraints of the form $y(a) = y_0$, $y(b) = y_1$. The length constraint is again given by (2.1). In the figure we took y_0 and y_1 to be equal, and also assumed that the suspension points are high enough and far enough apart compared to the length C_0 so that the chain does not touch the ground ($y(x) > 0$ for all x). The chain will take the shape of minimal potential energy. This amounts to saying that the y-coordinate of its center of mass should be minimized. Since the mass density is uniform, we integrate the y-coordinate of the point along the curve with respect to the arclength and obtain the functional

$$J(y) = \int_a^b y(x)\sqrt{1 + (y'(x))^2}dx$$

(the actual center of mass is $J(y)/C_0$). Thus the problem is to minimize this functional subject to the length constraint (2.1). As in Dido's problem, we need to assume that y is differentiable (at least almost everywhere); this time we do it also to ensure that the cost is well defined, because y' appears inside the integral in J.

The correct description of the catenary curve was obtained by Johann Bernoulli in 1691. It is given by the formula

$$y(x) = c\cosh(x/c), \qquad c > 0 \qquad (2.3)$$

(modulo a translation of the origin in the (x, y)-plane), unless the chain is suspended too low and touches the ground. The name "catenary" is derived from the Latin word *catena* (chain).

2.1.4   Brachistochrone

Given two fixed points in a vertical plane, we want to find a path between them such that a particle sliding without friction along this path takes the shortest possible time to travel from one point to the other (see Figure 2.4). The name "brachistochrone" was given to this problem by Johann Bernoulli; it comes from the Greek words βράχιστος (shortest) and χρόνος (time).

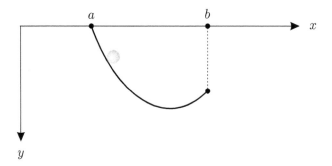

Figure 2.4: The brachistochrone problem

It is tempting to think that the solution is a straight line, but this is not the case. For example, if the two points are at the same height, then the particle will not move if placed on the horizontal line (we are assuming that the particle is initially at rest); on the other hand, traveling along a lower semicircle, the particle will reach the other point in finite time. Galileo, who studied this problem in 1638, showed that for points at different heights, it is also true that sliding down a straight line is slower than along an arc of a circle. Galileo thought that an arc of a circle is the optimal path, but this is not true either.

The travel time along a path is given by the integral of the ratio of the arclength to the particle's speed. We already know the expression for the arclength from (2.1). To obtain a formula for the speed, we use the law of conservation of energy. Let us choose coordinates as shown in Figure 2.4: the y-axis points downward and the x-axis passes through the initial point (the higher one). Then the kinetic and potential energy are both 0 initially, thus the total energy is always 0 and we have

$$\frac{mv^2}{2} - mgy = 0 \tag{2.4}$$

where v is the particle's speed, m is its mass, and g is the gravitational constant. Choosing suitable units, we can assume without loss of generality that $m = 1$ and $g = 1/2$. Then (2.4) simplifies to

$$v = \sqrt{y} \tag{2.5}$$

and so our problem becomes that of minimizing the value of the functional

$$J(y) = \int_a^b \frac{\sqrt{1 + (y'(x))^2}}{\sqrt{y(x)}} dx \qquad (2.6)$$

over all (almost everywhere) differentiable curves y connecting the two given points with x-coordinates a and b.

Johann Bernoulli posed the brachistochrone problem in 1696 as a challenge to his contemporaries. Besides Bernoulli himself, correct solutions were obtained by Leibniz, Newton, Johann's brother Jacob Bernoulli, and others. The optimal curves are *cycloids*, defined by the parametric equations

$$\begin{aligned} x(\theta) &= a + c(\theta - \sin\theta), \\ y(\theta) &= c(1 - \cos\theta) \end{aligned} \qquad (2.7)$$

where the parameter θ takes values between 0 and 2π and $c > 0$ is a constant. These equations describe the curve traced by a point on a circle of radius c as this circle rolls without slipping on the horizontal axis (see Figure 2.5). A somewhat surprising outcome is that the fastest motion may involve dipping below the desired height to build up speed and then coming back up (whether or not this happens depends on the relative location of the two points). This principle also applies to changing an airplane's altitude.

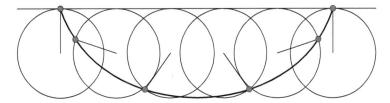

Figure 2.5: A cycloid

It is interesting to note that Johann Bernoulli's original solution to the brachistochrone problem was based on Snell's law (2.2) for light refraction. In view of equation (2.5), we can treat the particle as a light ray traveling in a medium where the speed of light is proportional to the square root of the height. In a discretized version of this situation, the vertical plane is divided into horizontal strips and the speed of light is constant in each strip. The light ray follows a straight line within each strip and bends at the boundaries according to Snell's law. From this piecewise linear path, a cycloid is obtained in the limit as the heights of the strips tend to 0. This solution can be obtained more easily (and more rigorously) using methods from calculus of variations and optimal control theory.

Exercise 2.1 *Find another example of a variational problem. Describe it verbally first, then formalize it by specifying admissible curves and giving* *an expression for the functional to be minimized or maximized. You do not need to solve it.*   □

2.2   BASIC CALCULUS OF VARIATIONS PROBLEM

The reader has probably observed that the problems of Dido, catenary, and brachistochrone, although different in their physical meaning, all take essentially the same mathematical form. We are now ready to turn to a general problem formulation that captures these examples and many related ones as special cases. For the moment we ignore the issue of constraints such as the arclength constraint (2.1), which was present in Dido's problem and the catenary problem; we will incorporate constraints of this type later (in Section 2.5).

The simplest version of the calculus of variations problem can be stated as follows. Consider a function $L : \mathbb{R} \times \mathbb{R} \times \mathbb{R} \to \mathbb{R}$.

Basic Calculus of Variations Problem: Among all $\mathcal{C}^1$ curves $y : [a, b] \to \mathbb{R}$ satisfying given boundary conditions

$$y(a) = y_0, \qquad y(b) = y_1 \tag{2.8}$$

find (local) minima of the cost functional

$$J(y) := \int_a^b L(x, y(x), y'(x))dx. \tag{2.9}$$

Since y takes values in $\mathbb{R}$, it represents a single planar curve connecting the two fixed points (a, y_0) and (b, y_1). This is the *single-degree-of-freedom* case. In the *multiple-degrees-of-freedom* case, one has $y : [a, b] \to \mathbb{R}^n$ and accordingly $L : \mathbb{R} \times \mathbb{R}^n \times \mathbb{R}^n \to \mathbb{R}$. This generalization is useful for treating spatial curves ($n = 3$) or for describing the motion of many particles; the latter setting was originally proposed by Lagrange in his 1788 monograph *Mécanique Analytique*. The assumption that $y \in \mathcal{C}^1$ is made to ensure that J is well defined (of course we do not need it if y' does not appear in L). We can allow y' to be discontinuous at some points; we will discuss such situations and see their importance soon.

The function L is called the *Lagrangian*, or the *running cost*. It is clear that a maximization problem can always be converted into a minimization problem by flipping the sign of L. In the analysis that follows, it will be important to remember the following point: Even though y and y' are the position and velocity along the curve, L is to be viewed as a function of three independent variables. To emphasize this fact, we will sometimes write $L = L(x, y, z)$. When deriving optimality conditions, we will need to impose some differentiability assumptions on L.

2.2.1 Weak and strong extrema

We recall from Section 1.3 that in order to define local optimality, we must first select a norm, and on the space of $\mathcal{C}^1$ curves $y : [a, b] \to \mathbb{R}$ there are two natural candidates for the norm: the 0-norm (1.29) and the 1-norm (1.30). Extrema (minima and maxima) of J with respect to the 0-norm are called *strong extrema*, and those with respect to the 1-norm are called *weak extrema*.

These two notions will be central to our subsequent developments, and so it is useful to reflect on them for a little while until the distinction between the two types of extrema becomes clear and there is no possibility of confusing them. If a $\mathcal{C}^1$ curve y^* is a strong extremum, then it is automatically a weak one, but the converse is not true. The reason is that an ε-ball around y^* with respect to the 0-norm contains the ε-ball with respect to the 1-norm for the same ε, as is clear from the norm definitions; on the other hand, the ε-ball with respect to the 1-norm does not contain the ε'-ball with respect to the 0-norm for any ε', no matter how small. In other words, it is harder to satisfy $J(y^*) \leq J(y)$ for all y close enough to y^* if we understand closeness in the sense of the 0-norm. Closeness in the sense of the 1-norm is a more restrictive condition, since the derivatives of y and y^* also have to be close, meaning that there are fewer perturbations to check than for the 0-norm. We will see that, for the same reason, studying weak extrema is easier than studying strong extrema.

On the other hand, it will become evident later that the concept of a weak minimum is not very suitable in optimal control. Indeed, an optimal trajectory y^* should give a lower cost than all nearby trajectories y, and there is no compelling reason to take into account the difference between the derivatives of y^* and y. Also, as we already mentioned, requiring y to be a $\mathcal{C}^1$ curve is often too restrictive. Specifically, we will want to allow curves y which are continuous everywhere on $[a, b]$ and whose derivative y' exists everywhere except possibly a finite number of points in $[a, b]$ and is continuous and bounded between these points. Let us agree to call such curves *piecewise $\mathcal{C}^1$*, to reflect the fact that they are concatenations of finitely many $\mathcal{C}^1$ pieces. (We could define the class of admissible curves more precisely using the notion of an absolutely continuous function; we will revisit this issue in Section 3.3 as we make the transition to optimal control.) If we use the 0-norm, then it makes no difference whether y is $\mathcal{C}^1$ or piecewise $\mathcal{C}^1$ or just $\mathcal{C}^0$; this is another advantage of the 0-norm over the 1-norm.

In view of the above remarks, it seems natural to first obtain some basic tools for studying weak minima and then proceed to develop more advanced tools for investigating strong minima. This is essentially what we will do. The next example illustrates some of the points that we just made regarding the 0-norm versus the 1-norm. The exercise that follows should help the reader to better grasp the concepts of weak and strong minima; it is to be solved using only the definitions.

Example 2.1   *Consider the three curves y_0, y_1, and y_2 shown in Figure 2.6. We think of y_1 and y_2 as results of perturbations of y_0 which are small in magnitude. Then y_1 is close to y_0 with respect to both the 0-norm and the 1-norm, but y_2 is close to y_0 only with respect to the 0-norm. Indeed, the 1-norm of $y_2 - y_0$ is large because the derivative of y_2 is large in magnitude due to the sharp spikes. (Technically speaking, to be able to take the 1-norm we should smoothen the corners of y_2.)*

Figure 2.6: Closeness in weak and strong sense

We can also consider another curve y_3 (not shown in the figure) which is of the same form as y_2 but slightly out of phase with it. Then y_2 and y_3 will be close with respect to the 0-norm but not the 1-norm. We can think of y_2 and y_3 as solutions of the control system $dy/dx = u \in \{-1, 1\}$. A small difference in phase amounts to a slight shift of the switching times of the control u. Such a small perturbation of the control should be admissible, which is why the 0-norm and strong extrema provide a more reasonable notion of local optimality.   □

Exercise 2.2   *Consider the problem of minimizing the functional*

$$J(y) = \int_0^1 (y'(x))^2 \left(1 - (y'(x))^2\right) dx$$

subject to the boundary conditions $y(0) = y(1) = 0$. Is the curve $y \equiv 0$ a weak minimum (over $\mathcal{C}^1$ curves)? Is it a strong minimum (over piecewise $\mathcal{C}^1$ curves)? Is there another curve that is a strong minimum?   □

2.3   FIRST-ORDER NECESSARY CONDITIONS FOR WEAK EXTREMA

In this section we will derive the most fundamental result in calculus of variations: the Euler-Lagrange equation. Unless stated otherwise, we will be working with the Basic Calculus of Variations Problem defined in Section 2.2. Thus our function space V is $\mathcal{C}^1([a, b], \mathbb{R})$, the subset A consists of functions $y \in V$ satisfying the boundary conditions (2.8), and the functional J to be minimized takes the form (2.9). The Euler-Lagrange equation provides a more explicit characterization of the first-order necessary condition (1.36) for this situation.

In deriving the Euler-Lagrange equation, we will follow the basic variational approach presented in Section 1.3.2 and consider nearby curves of the form

$$y + \alpha\eta \tag{2.10}$$

where the perturbation $\eta : [a, b] \to \mathbb{R}$ is another $\mathcal{C}^1$ curve and α varies in an interval around 0 in $\mathbb{R}$. For α close to 0, these perturbed curves are close to y in the sense of the 1-norm. For this reason, the resulting first-order necessary condition handles weak extrema. However, since a $\mathcal{C}^1$ strong extremum is automatically a weak extremum, every necessary condition for the latter is necessary for the former as well. Therefore, the Euler-Lagrange equation will apply to both weak and strong extrema, as long as we insist on working with $\mathcal{C}^1$ functions. As we know, using the 0-norm allows us to relax the $\mathcal{C}^1$ requirement, but this will in turn necessitate a different approach (i.e., a refined perturbation family) when developing optimality conditions. We will thus refer to the conditions derived in this section as necessary conditions for weak extrema, in order to distinguish them from sharper conditions to be given later which apply specifically to strong extrema. Similar remarks apply to other necessary conditions to be derived in this chapter. The sufficient condition of Section 2.6.2, on the other hand, will apply to weak minima only.

2.3.1   Euler-Lagrange equation

We continue to follow the notational convention of Chapter 1 and denote by L_x, L_y, L_z, L_{xx}, L_{xy}, etc. the partial derivatives of the Lagrangian $L = L(x, y, z)$. To keep things simple, we assume that all derivatives appearing in our calculations exist and are continuous. While we will not focus on spelling out the weakest possible regularity assumptions on L, we will make some remarks to clarify this issue in Section 2.3.3.

Let $y = y(x)$ be a given test curve in A. For a perturbation η in (2.10) to be admissible, the new curve (2.10) must again satisfy the boundary conditions (2.8). Clearly, this is true if and only if

$$\eta(a) = \eta(b) = 0. \tag{2.11}$$

In other words, we must only consider perturbations vanishing at the endpoints. Now, the first-order necessary condition (1.36) says that if y is a local extremum of J, then for every η satisfying (2.11) we must have $\delta J|_y (\eta) = 0$. (We denoted extrema by y^* in (1.36) but here we drop the asterisks to avoid overly cluttered notation.) In the present case we want to go further and use the specific form of J given by (2.9) to arrive at a more explicit condition in terms of the Lagrangian L.

Recall that the first variation $\delta J|_y$ was defined via

$$J(y + \alpha \eta) = J(y) + \delta J|_y (\eta)\alpha + o(\alpha). \tag{2.12}$$

The left-hand side of (2.12) is

$$J(y + \alpha \eta) = \int_a^b L(x, y(x) + \alpha \eta(x), y'(x) + \alpha \eta'(x))dx. \tag{2.13}$$

We can write down its first-order Taylor expansion with respect to α by expanding the expression inside the integral with the help of the chain rule:

$$J(y + \alpha \eta) = \int_a^b \Big(L(x, y(x), y'(x)) + L_y(x, y(x), y'(x))\alpha \eta(x)$$
$$+ L_z(x, y(x), y'(x))\alpha \eta'(x) + o(\alpha) \Big)dx.$$

Matching this with the right-hand side of (2.12), we deduce that the first variation is

$$\delta J|_y (\eta) = \int_a^b \Big(L_y(x, y(x), y'(x))\eta(x) + L_z(x, y(x), y'(x))\eta'(x) \Big)dx. \tag{2.14}$$

Note that, proceeding slightly differently, we could arrive at the same result by remembering from (1.33)–(1.35) that

$$\delta J|_y (\eta) = \lim_{\alpha \to 0} \frac{J(y + \alpha \eta) - J(y)}{\alpha} = \frac{d}{d\alpha}\Big|_{\alpha=0} J(y + \alpha \eta)$$

and using differentiation under the integral sign on the right-hand side of (2.13).

Exercise 2.3   *Prove that $\delta J|_y$ given by (2.14) also satisfies the alternative definition of the first variation based on (1.37) in which the norm is the 1-norm. Is this true for the 0-norm as well?*   □

We see that the first variation depends not just on η but also on η'. This is not surprising since L has y' as one of its arguments. However, we can eliminate the dependence on η' if we apply integration by parts to the second term on the right-hand side of (2.14):

$$\delta J|_y (\eta) = \int_a^b \Big(L_y(x, y(x), y'(x))\eta(x) - \eta(x)\frac{d}{dx}L_z(x, y(x), y'(x)) \Big)dx$$
$$+ L_z(x, y(x), y'(x))\eta(x)\big|_a^b \tag{2.15}$$

where the last term is 0 when η satisfies the boundary conditions (2.11). Thus we conclude that if y is an extremum, then we must have

$$\int_a^b \left(L_y(x, y(x), y'(x)) - \frac{d}{dx}L_z(x, y(x), y'(x))\right)\eta(x)dx = 0 \qquad (2.16)$$

for all $\mathcal{C}^1$ curves η vanishing at the endpoints $x = a$ and $x = b$.

The condition (2.16) does not yet give us a practically useful test for optimality, because we would need to check it for all admissible perturbations η. However, it is logical to suspect that the only way (2.16) can hold is if the term inside the parentheses—which does not depend on η—equals 0 for all x. The next lemma shows that this is indeed the case.

LEMMA 2.1. *If a continuous function* $\xi : [a, b] \to \mathbb{R}$ *is such that*

$$\int_a^b \xi(x)\eta(x)dx = 0$$

for all $\mathcal{C}^1$ *functions* $\eta : [a, b] \to \mathbb{R}$ *with* $\eta(a) = \eta(b) = 0$, *then* $\xi \equiv 0$.

PROOF. Suppose that $\xi(\bar{x}) \neq 0$ for some $\bar{x} \in [a, b]$. By continuity, ξ is then nonzero and maintains the same sign on some subinterval $[c, d]$ containing $\bar{x}$. Just for concreteness, let us say that ξ is positive on $[c, d]$.

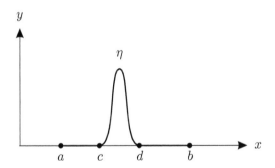

Figure 2.7: The graph of η

Construct a function $\eta \in \mathcal{C}^1([a, b], \mathbb{R})$ that is positive on (c, d) and 0 everywhere else (see Figure 2.7). For example, we can set $\eta(x) = (x-c)^2(x-d)^2$ for $x \in [c, d]$ and $\eta(x) = 0$ otherwise. This gives $\int_a^b \xi(x)\eta(x)dx > 0$, and we reach a contradiction. $\square$

It follows from (2.16) and Lemma 2.1 that for $y(\cdot)$ to be an extremum, a necessary condition is

$$L_y(x, y(x), y'(x)) = \frac{d}{dx}L_z(x, y(x), y'(x)) \qquad \forall\, x \in [a, b]. \qquad (2.17)$$

This is the celebrated **Euler-Lagrange equation** providing the **first-order necessary condition for optimality**. It is often written in the shorter form

$$\boxed{L_y = \frac{d}{dx}L_{y'}}$$

(2.18)

We must keep in mind, however, that the correct interpretation of the Euler-Lagrange equation is (2.17): y and y' are treated as independent variables when computing the partial derivatives L_y and $L_{y'}$, then one plugs in for these variables the position $y(x)$ and velocity $y'(x)$ of the curve, and finally the differentiation with respect to x is performed using the chain rule. Written out in detail, the right-hand side of (2.17) is

$$\frac{d}{dx}L_z(x, y(x), y'(x)) = L_{zx}(x, y(x), y'(x)) + L_{zy}(x, y(x), y'(x))y'(x)$$
$$+ L_{zz}(x, y(x), y'(x))y''(x).$$

(2.19)

It might not be necessary to actually perform all of these operations, though, as the next example demonstrates.

Example 2.2   *Let us find the shortest path between two points in the plane. Of course the answer is obvious, but let us see how we can obtain it from the Euler-Lagrange equation. We already discussed the functional that describes the length of a curve in Section 2.1, in the context of Dido's problem and the catenary problem (where it played the role of a side constraint rather than the cost functional). This length functional is $J(y) = \int_a^b \sqrt{1 + (y'(x))^2}dx$, hence the Lagrangian is $L(x, y, z) = \sqrt{1 + z^2}$. Since $L_y = 0$, the Euler-Lagrange equation reads $\frac{d}{dx}L_z(x, y(x), y'(x)) = 0$. This implies that L_z stays constant along the shortest path. We have $L_z = z/\sqrt{1 + z^2}$, thus*

$$L_z(x, y(x), y'(x)) = \frac{y'(x)}{\sqrt{1 + (y'(x))^2}}.$$

(2.20)

For this to be constant, y' must be constant, i.e., the path must be a straight line. The unique straight line connecting two given points is clearly the shortest path between them. Note that we did not need to compute the derivative $\frac{d}{dx}L_z(x, y(x), y'(x))$.   □

The functional J to be minimized is given by the integral of the Lagrangian L along a path, while the Euler-Lagrange equation involves derivatives of L and must hold for every point on the optimal path; observe that the integral has disappeared. The underlying reason is that if a path is optimal, then every infinitesimally small portion of it is optimal as well (no "shortcuts" are possible). The exact mechanism by which we pass from the statement that the integral is minimized to the pointwise condition is revealed by Lemma 2.1 and its proof.

Trajectories satisfying the Euler-Lagrange equation (2.18) are called *extremals* (of the functional J). Since the Euler-Lagrange equation is only a necessary condition for optimality, not every extremal is an extremum. We see from (2.19) that the equation (2.18) is a second-order differential equation; thus we expect that generically, the two boundary conditions (2.8) should be enough to specify a unique extremal. When this is true—as is the case in the above example—and when an optimal curve is known to exist, we can actually conclude that the unique extremal gives the optimal solution. In general, the question of existence of optimal solutions is not trivial, and the following example should serve as a warning. (We will come back to this issue later in the context of optimal control.)

Example 2.3   *Consider the problem of minimizing the functional $J(y) = \int_0^1 y(x)(y'(x))^2 dx$ subject to the boundary conditions $y(0) = y(1) = 0$. The Euler-Lagrange equation is $\frac{d}{dx}(2yy') = (y')^2$, and $y \equiv 0$ is a solution. Actually, one can show that this is a unique extremal satisfying the boundary conditions (we leave the proof of this fact to the reader). But $y \equiv 0$ is easily seen to be neither a minimum nor a maximum.* $\square$

2.3.2   Historical remarks

The equation (2.18) was derived by Euler around 1740. His original derivation was very different from the one we gave, and relied on discretization. The idea is to approximate a general curve by a piecewise linear one passing through N points, as in Figure 2.8. The problem of optimizing the locations of these N points is finite-dimensional, hence it can be solved using standard calculus. The equation (2.18) is obtained in the limit as $N \to \infty$.

Figure 2.8: Illustrating Euler's derivation

An alternative way to arrive at the same result, free of geometric considerations and relying on analysis alone, was proposed by Lagrange in 1755 (when he was only 19 years old). Lagrange described his argument in a letter to Euler, who was quite impressed by it and subsequently coined the term "calculus of variations" for Lagrange's method. Even though the problem formulation and the solution given by Lagrange differ from the modern treatment in the notation and other details, the main ideas behind our derivation of the Euler-Lagrange equation are essentially contained in his work.

2.3.3 Technical remarks

It is straightforward to extend the necessary condition (2.18) to the multiple-degrees-of-freedom setting, in which $y = (y_1, \ldots, y_n)^T \in \mathbb{R}^n$. The same derivation is valid, provided that L_y and L_z are interpreted as gradient vectors of L with respect to y and z, respectively, and by products of vector quantities (such as $L_y \cdot \eta$) one means inner products in $\mathbb{R}^n$. The result is the same Euler-Lagrange equation (2.18), which is now perhaps easier to apply and interpret if it is written componentwise:

$$L_{y_i} = \frac{d}{dx} L_{y_i'}, \qquad i = 1, \ldots, n. \tag{2.21}$$

Returning to the single-degree-of-freedom case, let us examine more carefully under what differentiability assumptions our derivation of the Euler-Lagrange equation is valid. We applied Lemma 2.1 to the function ξ given by the expression inside the large parentheses in (2.16), whose second term is shown in more detail in (2.19). In the lemma, ξ must be continuous. The appearance of second-order partial derivatives of L in (2.19) suggests that we should assume $L \in \mathcal{C}^2$. Somewhat more alarmingly, the presence of the term $L_{zz}y''$ indicates that we should assume $y \in \mathcal{C}^2$, and not merely $y \in \mathcal{C}^1$ as in our original formulation of the Basic Calculus of Variations Problem. Fortunately, we can avoid making this assumption if we proceed more carefully, as follows. Let us apply integration by parts to the first term rather than the second term on the right-hand side of (2.14):

$$\delta J|_y(\eta) = \int_a^b \left(L_z(x, y(x), y'(x))\eta'(x) - \eta'(x) \int_a^x L_y(w, y(w), y'(w))dw \right) dx$$
$$+ \left. \eta(x) \int_a^x L_y(w, y(w), y'(w))dw \right|_a^b$$

where the last term again vanishes for our class of perturbations η. Thus an extremum y must satisfy

$$\int_a^b \left(L_z(x, y(x), y'(x)) - \int_a^x L_y(w, y(w), y'(w))dw \right) \eta'(x)dx = 0. \tag{2.22}$$

We now need the following modification of Lemma 2.1.

LEMMA 2.2. *If a continuous function $\xi : [a, b] \to \mathbb{R}$ is such that*

$$\int_a^b \xi(x)\eta'(x)dx = 0$$

for all $\mathcal{C}^1$ functions $\eta : [a, b] \to \mathbb{R}$ with $\eta(a) = \eta(b) = 0$, then ξ is a constant function.

Exercise 2.4   *Prove Lemma 2.2.*   □

From (2.22) and Lemma 2.2 we obtain that along an optimal curve we must have

$$L_z(x, y(x), y'(x)) = \int_a^x L_y(w, y(w), y'(w))dw + C \qquad (2.23)$$

where C is a constant. It follows that the derivative $\frac{d}{dx}L_z(x, y(x), y'(x))$ indeed exists and equals $L_y(x, y(x), y'(x))$, and we do not need to make extra assumptions to guarantee the existence of this derivative. In other words, curves on which the first variation vanishes automatically enjoy additional regularity. Thus for the necessary condition described by the Euler-Lagrange equation to be valid, it is enough to assume that $y \in \mathcal{C}^1$ and $L \in \mathcal{C}^1$. The integral form (2.23) of the Euler-Lagrange equation actually applies to extrema over piecewise $\mathcal{C}^1$ curves as well, although we would need a more general version of Lemma 2.2 to establish this fact.[2] The $\mathcal{C}^1$ assumption on L can be further relaxed; note, in particular, that the partial derivative L_x did not appear anywhere in our analysis.

It can be shown that the Euler-Lagrange equation is invariant under arbitrary changes of coordinates, i.e., it takes the same form in every coordinate system. This allows one to pick convenient coordinates when studying a specific problem. For example, in certain mechanical problems it is natural to use polar coordinates; we will come across such a case in Section 2.4.3.

2.3.4   Two special cases

We know from the discussion given towards the end of Section 2.3.1—see the formula (2.19) in particular—that the Euler-Lagrange equation (2.18) is a second-order differential equation for $y(\cdot)$; indeed, it can be written in more detail as

$$L_y = L_{y'x} + L_{y'y}y' + L_{y'y'}y''. \qquad (2.24)$$

Note that we are being somewhat informal in denoting the third argument of L by y', and also in omitting the x-arguments. We now discuss two special cases in which (2.24) can be reduced to a differential equation that is of first order, and therefore more tractable.

---

[2]Since we are discussing weak extrema, to accommodate piecewise $\mathcal{C}^1$ curves we need to slightly generalize the definition of the 1-norm; see Section 3.1.1 for details. Also, the equation (2.23) holds away from the discontinuities of y'. To prove this, we would need to modify the proof of Lemma 2.2 to cover functions ξ that are only piecewise continuous (see page 84 for a precise definition of this class), while still working with $\eta \in \mathcal{C}^1$.

SPECIAL CASE 1 ("no y"). This refers to the situation where the Lagrangian does not depend on y, i.e., $L = L(x, y')$. The Euler-Lagrange equation (2.18) becomes $\frac{d}{dx} L_{y'} = 0$, which means that $L_{y'}$ must stay constant. In other words, extremals are solutions of the first-order differential equation

$$L_{y'}(x, y'(x)) = c \qquad (2.25)$$

for various values of $c \in \mathbb{R}$. We already encountered such a situation in Example 2.2, where we actually had $L = L(y')$. Due to the presence of the parameter c, we expect that the family of solutions of (2.25) is rich enough to contain one (and only one) extremal that passes through two given points.

The quantity $L_{y'}$, evaluated along a given curve, is called the *momentum*. This terminology will be justified in Section 2.4.

SPECIAL CASE 2 ("no x"). Suppose now that the Lagrangian does not depend on x, i.e., $L = L(y, y')$. In this case the partial derivative $L_{y'x}$ vanishes from (2.24), and the Euler-Lagrange equation becomes

$$0 = L_{y'y} y' + L_{y'y'} y'' - L_y.$$

Multiplying both sides by y', we have

$$0 = L_{y'y}(y')^2 + L_{y'y'} y' y'' - L_y y' = \frac{d}{dx}\left(L_{y'} y' - L \right)$$

where the last equality is easily verified (the $L_{y'} y''$ terms cancel out). This means that $L_{y'} y' - L$ must remain constant. Thus, similarly to Case 1, extremals are described by the family of first-order differential equations

$$L_{y'}(y(x), y'(x)) y'(x) - L(y(x), y'(x)) = c$$

parameterized by $c \in \mathbb{R}$.

The quantity $L_{y'} y' - L$ is called the *Hamiltonian*. Although its significance is not yet clear at this point, it will play a crucial role throughout the rest of the book.

Exercise 2.5   *Use the above "no x" result to show that extremals for the brachistochrone problem are indeed given by the equations (2.7).*   □

2.3.5   Variable-endpoint problems

We know that the first-order necessary condition (1.36)—which serves as the basis for the Euler-Lagrange equation—need only hold for admissible perturbations. So far we have been considering the Basic Calculus of Variations Problem, in which the curves have both their endpoints fixed by the boundary conditions (2.8). Accordingly, the class of admissible perturbations is restricted to those vanishing at the endpoints. This fact, reflected in (2.11), was explicitly used in the derivation of the Euler-Lagrange equation (2.18).

If we change the boundary conditions for the curves of interest, then the class of admissible perturbations will also change, and in general the necessary condition for optimality will be different. To give an example of such a situation, we now consider a simple variable-endpoint problem. Suppose that the cost functional takes the same form (2.9) as before, the initial point of the curve is still fixed by the boundary condition $y(a) = y_0$, but the terminal point $y(b)$ is free. The resulting family of curves is depicted in Figure 2.9.

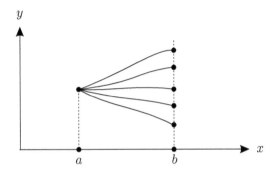

Figure 2.9: Variable terminal point

The perturbations η must still satisfy $\eta(a) = 0$ but $\eta(b)$ can be arbitrary. In view of (2.15), the first variation is then given by

$$
\delta J|_y (\eta) = \int_a^b \left(L_y(x, y(x), y'(x)) - \frac{d}{dx} L_z(x, y(x), y'(x)) \right) \eta(x) dx \\
+ L_z(b, y(b), y'(b)) \eta(b)
\tag{2.26}
$$

and this must be 0 if y is to be an extremum. Perturbations such that $\eta(b) = 0$ are still allowed; let us consider them first. They make the last term in (2.26) disappear, leaving us with (2.16). Exactly as before, we deduce from this that the Euler-Lagrange equation must hold, i.e., it is still a necessary condition for optimality. The Euler-Lagrange equation says that the expression in the large parentheses inside the integral in (2.26) is 0. But this means that the entire integral is 0, for *all* admissible η (not just those vanishing at $x = b$). The last term in (2.26) must then also vanish, which gives us an additional necessary condition for optimality:

$$
L_z(b, y(b), y'(b)) \eta(b) = 0
$$

or, since $\eta(b)$ is arbitrary,

$$
L_z(b, y(b), y'(b)) = 0.
\tag{2.27}
$$

We can think of (2.27) as replacing the boundary condition $y(b) = y_1$. Recall that we want to have two boundary conditions to uniquely specify an extremal. Comparing with the Basic Calculus of Variations Problem, here we have only one endpoint fixed a priori, but on the other hand we have a richer perturbation family which allows us to obtain one extra condition (2.27).

Example 2.4   *Consider again the length functional from Example 2.2, with Lagrangian $L(x, y, z) = \sqrt{1 + z^2}$. In other words, we are looking for a shortest path from a given point to a vertical line. Using the formula (2.20), we see that the condition (2.27) amounts to $y'(b) = 0$. This means that the optimal path must have a horizontal tangent at the final point (i.e., it must meet the vertical line of possible final points orthogonally). We already know that in order to satisfy the Euler-Lagrange equation, the path must be a straight line. Thus the only path satisfying the necessary conditions is a horizontal line, which is of course the optimal solution.* ☐

Exercise 2.6   *Consider a more general version of the above variable-terminal-point problem, with the vertical line replaced by a curve:*

$$J(y) = \int_a^{x_f} L(x, y(x), y'(x)) dx$$

where $y(a) = y_0$ is fixed, x_f is unspecified, and $y(x_f) = \varphi(x_f)$ for a given C^1 function $\varphi : \mathbb{R} \to \mathbb{R}$. Derive a necessary condition for a weak extremum. Your answer should contain, besides the Euler-Lagrange equation, an additional condition ("transversality condition") which accounts for variations in x_f and which explicitly involves φ'. ☐

Working on this exercise, the reader will realize that obtaining a transversality condition in the specified form requires a somewhat more advanced analysis than what we have done so far. We will employ similar techniques again soon when deriving conditions for strong minima in Section 3.1.1. The transversality condition itself is essentially a preview of what we will see later in the context of the maximum principle. More general variable-endpoint problems in which the initial point is allowed to vary as well, and the resulting transversality conditions, will also be mentioned in the optimal control setting (at the end of Section 4.3).

2.4   HAMILTONIAN FORMALISM AND MECHANICS

We now present an alternative formulation of the results of Euler and Lagrange, proposed many years later by Hamilton. These mathematical developments will be of great significance to us later in the context of optimal control. We will also discuss the physical interpretation of the relevant concepts and equations, which is somewhat secondary to our main goals but nevertheless interesting and important.

2.4.1   Hamilton's canonical equations

In Section 2.3.4 we came across two quantities, which we recall here and for which we now introduce special symbols. The first one was the *momentum*

$$p := L_{y'}(x, y, y') \tag{2.28}$$

which we will usually regard as a function of x associated to a given curve $y = y(x)$. The second object was the *Hamiltonian*

$$H(x, y, y', p) := p \cdot y' - L(x, y, y') \tag{2.29}$$

which is written here as a general function of four variables but also becomes a function of x alone when evaluated along a curve. The inner product sign $\cdot$ in the definition of H reflects the fact that in the multiple-degrees-of-freedom case, y' and p are vectors.

The variables y and p are called the *canonical variables*. Suppose now that y is an extremal, i.e., satisfies the Euler-Lagrange equation (2.18). It turns out that the differential equations describing the evolution of y and p along such a curve, when written in terms of the Hamiltonian H, take a particularly nice form. For y, we have

$$\frac{dy}{dx} = y'(x) = H_p(x, y(x), y'(x)).$$

For p, we have

$$\frac{dp}{dx} = \frac{d}{dx} L_{y'}(x, y(x), y'(x)) = L_y(x, y(x), y'(x)) = -H_y(x, y(x), y'(x))$$

where the second equality is the Euler-Lagrange equation. In more concise form, the result is

$$\boxed{y' = H_p, \quad p' = -H_y} \tag{2.30}$$

which is known as the system of *Hamilton's canonical equations*. This reformulation of the Euler-Lagrange equation was proposed by Hamilton in 1835. Since we are not assuming here that we are in the "no y" case or the "no x" case of Section 2.3.4, the momentum p and the Hamiltonian H need not be constant along extremals.

Exercise 2.7 *Confirm directly from the equations (2.28)–(2.30) that in the "no y" case p is constant along extremals and in the "no x" case H is constant along extremals.* □

An important additional observation is that the partial derivative of H with respect to y' is

$$H_{y'}(x, y, y', p) = p - L_{y'}(x, y, y') = 0 \qquad (2.31)$$

where the last equality follows from the definition (2.28) of p. This suggests that, in addition to the canonical equations (2.30), another necessary condition for optimality should be that H has a stationary point as a function of y' along an optimal curve. To make this statement more precise, let us plug the following arguments into the Hamiltonian: an arbitrary $x \in [a, b]$; for y, the corresponding position $y(x)$ of the optimal curve; for p, the corresponding value of the momentum $p(x) = L_{y'}(x, y(x), y'(x))$. Let us keep the last remaining argument, y', as a free variable, and relabel it as z for clarity. This yields the function

$$H^*(z) := L_{y'}(x, y(x), y'(x)) \cdot z - L(x, y(x), z). \qquad (2.32)$$

Our claim is that this function has a stationary point when z equals $y'(x)$, the velocity of the optimal curve at x. Indeed, it is immediate to check that

$$\frac{dH^*}{dz}(y'(x)) = 0. \qquad (2.33)$$

Later we will see that in the context of the maximum principle this stationary point is actually an extremum, in fact, a maximum. Moreover, the statement about the maximum remains true when H is not necessarily differentiable or when y' takes values in a set with a boundary and $H_{y'} \neq 0$ on this boundary; the basic property is not that the derivative vanishes but that H achieves the maximum in the above sense.

Mathematically, the Lagrangian L and the Hamiltonian H are related via a construction known as the *Legendre transformation*. Since this transformation is classical and finds applications in many diverse areas (optimization, geometry, physics), we now proceed to describe it. However, we will see that it does not quite provide the right point of view for our future developments, and it is included here mainly for historical reasons.

2.4.2   Legendre transformation

Consider a function $f : \mathbb{R} \to \mathbb{R}$, whose argument we denote by ξ (the curve in Figure 2.10 is a possible graph of f). For simplicity we are considering the scalar case, but the extension to $f : \mathbb{R}^n \to \mathbb{R}$ is straightforward. The *Legendre transform* of f will be a new function, f^*, of a new variable, $p \in \mathbb{R}$.

Let p be given. Draw a line through the origin with slope p. Take a point $\xi = \xi(p)$ at which the (directed) vertical distance from the graph of f to this line is maximized:

$$\xi(p) := \arg\max_{\xi}\{p\xi - f(\xi)\}. \qquad (2.34)$$

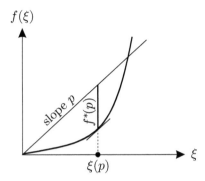

Figure 2.10: Legendre transformation

(Note that $\xi(p)$ may not exist, so the domain of f^* is not known a priori. Also, $\xi(p)$ is not necessarily unique unless f is a strictly convex function.) Now, define $f^*(p)$ to be this maximal value of the gap between $p\xi$ and $f(\xi)$:

$$f^*(p) := p\xi(p) - f(\xi(p)) = \max_{\xi}\{p\xi - f(\xi)\}. \tag{2.35}$$

We can also write this definition more symmetrically as

$$f^*(p) + f(\xi) = p\xi \tag{2.36}$$

where p and $\xi = \xi(p)$ are related via (2.34). When f is differentiable, the maximization condition (2.34) implies that the derivative of $p\xi - f(\xi)$ with respect to ξ must equal 0 at $\xi(p)$:

$$p - f'(\xi(p)) = 0. \tag{2.37}$$

Geometrically, the tangent line to the graph of f at $\xi(p)$ must have slope p, i.e., it must be parallel to the original line through the origin (see Figure 2.10). If f is convex then (2.34) and (2.37) are equivalent.

The Legendre transformation has some nice properties. For example, f^* is a convex function even if f is not convex. The reason is that f^* is a pointwise maximum of functions that are affine in p, as is clear from (2.35). Also, for convex functions the Legendre transformation is involutive: if f is convex, then $f^{**} = f$.

Now let us return to the Hamiltonian H defined in (2.29). We claim that it can be obtained by applying the Legendre transformation to the Lagrangian L. More precisely, for arbitrary fixed x and y let us consider $L(x, y, y')$ as a function of $\xi = y'$. The relation (2.37) between p and $\xi(p) = y'(p)$ becomes

$$p - L_{y'}(x, y, y'(p)) = 0 \tag{2.38}$$

which corresponds to our earlier definition (2.28) of the momentum p. Next, (2.35) gives

$$L^*(x, y, p) = py'(p) - L(x, y, y'(p)) \tag{2.39}$$

which is essentially our earlier definition (2.29) of the Hamiltonian H. But there is a difference: in (2.29) we had y' as an independent argument of H, while in (2.39) y' is a dependent variable expressed in terms of x, y, p by the implicit relation (2.38). In other words, the Legendre transform of $L(x, y, y')$ as a function of y' (with x, y fixed) is $H(x, y, p)$, which is a function of p (with x, y fixed) and no longer has y' as an argument. Note that the above derivation is formal, i.e., we are ignoring the question of whether or not (2.38) can indeed be solved for y'. This issue did not arise earlier when we were working with the Hamiltonian $H = H(x, y, y', p)$.

The above approach has another, more important drawback. Recall the observation based on (2.31) that H has a stationary point as a function of y' along an optimal curve. This property will be crucial later; combined with the canonical equations (2.30), it will lead us to the maximum principle. But it only makes sense when we treat y' as an independent variable in the definition of H. On the other hand, Hamilton and other 19th century mathematicians did not write the Hamiltonian in this way; they followed the convention of viewing y' as a dependent variable defined implicitly by (2.38). This is probably why it was not until the late 1950s that the maximum principle was discovered.

2.4.3   Principle of least action and conservation laws

Newton's second law of motion in the three-dimensional space can be written as the vector equation

$$\frac{d}{dt}(m\dot{q}) = -U_q \tag{2.40}$$

where $q = (x, y, z)^T$ is the vector of coordinates, $\dot{q} = dq/dt$ is the velocity vector, and $U = U(q)$ is the potential; consequently, $m\dot{q}$ is the momentum and $-U_q$ is the force. Note that we are only considering situations where the force is conservative, i.e., corresponds to the negative gradient of some potential function. Planar motion is obtained as a special case by dropping the z-coordinate.

It turns out that there is a direct relationship between (2.40) and the Euler-Lagrange equation. This is difficult to see right now because the notation in (2.40) is very different from the one we have been using. So, let us modify our earlier notation to better match (2.40). First, let us write t instead of x for the independent variable. Second, let us write q instead of y for the dependent variable. Then also y' becomes $\dot{q}$ and we have $L = L(t, q, \dot{q})$. In the new notation, the Euler-Lagrange equation becomes

$$\frac{d}{dt}L_{\dot{q}} = L_q. \tag{2.41}$$

Note that since $q \in \mathbb{R}^3$, we are referring to the multiple-degrees-of-freedom version (2.21) of the Euler-Lagrange equation, with $n = 3$.

Some remarks on the above change of notation are in order (as it is also a preview of things to come). The change from x to t is conceptually significant, because it implies that the curves are parameterized by *time* and thus describe some dynamic behavior (e.g., trajectories of a moving object). In problems such as Dido's problem or catenary problem, where there is no motion with respect to time, this notation would not be justified. Other variational problems, such as the brachistochrone problem, do indeed deal with paths of a moving particle. (We did not, however, explicitly use time when formulating the brachistochrone problem, and it would not make sense to just relabel x as t in our earlier formulation of that problem. Instead, we would need to reparameterize the (x, y)-trajectories with respect to time, which yields a different Lagrangian; we will do this in Section 3.2.) In mechanics, as well as in control theory, time is the default independent variable. Accordingly, when we come to the optimal control part of the book, we will make this change of notation permanent. For now, we adopt it just temporarily while we discuss applications of calculus of variations in mechanics. As for the (time-)dependent coordinate variables, it is natural to denote them by x and y for planar curves and by x, y and z for spatial curves. In general, the selection of a label for the coordinate vector is just a matter of preference and convention; the mechanics literature typically favors q or r, while in control theory it is customary to use x.

Let us now compare (2.41) with (2.40). Is there a choice of the Lagrangian L that would make these two equations the same? The answer is yes, and the reader will have no difficulty in seeing that the following Lagrangian does the job:

$$L := \frac{1}{2}m\left(\dot{x}^2 + \dot{y}^2 + \dot{z}^2\right) - U(q) \qquad (2.42)$$

which is the difference between the kinetic energy $T := \frac{1}{2}m(\dot{q} \cdot \dot{q})$ and the potential energy. We conclude that Newton's equations of motion can be recovered from a path optimization problem. This important result is known as **Hamilton's principle of least action**: *Trajectories of mechanical systems are extremals of the functional*

$$\int_{t_0}^{t_1} (T - U)dt$$

which is called the *action integral*. In general, these extremals are not necessarily minima. However, they are indeed minima—hence the term *"least action"* is accurate—if the time horizon is sufficiently short; this will follow from the second-order sufficient condition for optimality, to be derived in Section 2.6.2.

For example, if the potential is 0, then the trajectories are the extremals of the functional $\int_{t_0}^{t_1} \frac{1}{2}m(\dot{q} \cdot \dot{q})dt$, which are straight lines. We saw in Example 2.2 that straight lines arise as extremals when the Lagrangian is the arclength; the kinetic energy gives the same extremals. In the presence of gravity, the paths along which the action integral is minimized can be viewed as "straight lines" (shortest paths, or geodesics) in a curved space whose metric is determined by gravitational forces. This view of mechanics forms the basis for Einstein's theory of general relativity.

Observe the difference between the law that the derivative of the momentum equals the force and the principle that the action integral is minimized. The former condition holds pointwise in time, while the latter is a statement about the entire trajectory. However, their relation is not surprising, because if the action integral is minimized then every small piece of the trajectory must also deliver minimal action. In the limit as the length of the piece approaches 0, we recover the differential statement. We already discussed this point in the general context of the Euler-Lagrange equation (see page 38).

Now, what is the physical meaning of the Hamiltonian? Substituting q for y in (2.28)–(2.29) and using (2.42), we have

$$H = L_{\dot{q}} \cdot \dot{q} - L = \frac{1}{2}m(\dot{q} \cdot \dot{q}) + U = T + U = E$$

which is the total energy (kinetic plus potential). We see that the Hamiltonian not only enables a convenient rewriting of the Euler-Lagrange equation in the form of the canonical equations (2.30), but also has a very clear mechanical interpretation.

Exercise 2.8 *Reproduce all the derivations of this subsection (up to this point) for the more general setting of N moving particles.*   □

Recall that in Section 2.3.4 we studied two special cases for which we found conserved quantities, i.e., functions that remain constant along extremals. We now revisit these two *conservation laws*—as well as another related case—in the context of the principle of least action, which permits us to see their physical meaning.

CONSERVATION OF ENERGY. In a conservative system, the potential is fixed and does not change with time. Since the kinetic energy does not explicitly depend on time either, we have $L = L(q, \dot{q})$. In other words, the Lagrangian is invariant under time shifting. In our old notation, this corresponds to the "no x" case from Section 2.3.4, and we know that in this case the Hamiltonian is conserved. We also just saw that the Hamiltonian is the total energy of the system. Therefore, this is nothing but the well-known principle of conservation of energy.

CONSERVATION OF MOMENTUM. Suppose that no force is acting on the system (i.e., the system is closed). Since the force is given by $L_q = -U_q$, this implies that U must be constant. The kinetic energy T depends on $\dot{q}$ but not on q. Thus the Lagrangian $L = T - U$ does not explicitly depend on q, which means that it is invariant under parallel translations. This situation corresponds to the "no y" case from Section 2.3.4, where we saw that the *momentum*, $L_{\dot{q}} = m\dot{q}$ in the present notation, is conserved. A more general statement is that for each coordinate q_i that does not appear in L, the corresponding component $L_{\dot{q}_i} = m\dot{q}_i$ of the momentum is conserved.

CONSERVATION OF ANGULAR MOMENTUM. Consider a planar motion in a central field; in polar coordinates (r, θ), this is defined by the property that $U = U(r)$, i.e., the potential depends only on the radius and not on the angle. This means that no *torque* is acting on the system, making the Lagrangian L invariant under rotations. Now we can use the fact (noted at the end of Section 2.3.3) that the Euler-Lagrange equation looks the same in all coordinate systems. In particular, in polar coordinates we have

$$L_\theta = \frac{d}{dt}L_{\dot{\theta}} \tag{2.43}$$

and the analogous equation for r (which we do not need here). Arguing exactly as before, we can show that in the present "no θ" case the corresponding component of the momentum, $L_{\dot{\theta}}$, is conserved. Converting the kinetic energy from Cartesian to polar coordinates, we have

$$T = \frac{1}{2}m(\dot{x}^2 + \dot{y}^2) = \frac{1}{2}m(\dot{r}^2 + r^2\dot{\theta}^2). \tag{2.44}$$

Thus the conserved quantity, in polar and Cartesian coordinates, is

$$L_{\dot{\theta}} = T_{\dot{\theta}} = mr^2\dot{\theta} = m(x\dot{y} - y\dot{x}).$$

These are familiar expressions for the *angular momentum*.

We remark that all of the above examples are special instances of a general result known as Noether's theorem, which says that invariance of the action integral under some transformation (e.g., time shift, translation, rotation) implies the existence of a conserved quantity.

2.5 VARIATIONAL PROBLEMS WITH CONSTRAINTS

In Section 2.3 we showed that the Euler-Lagrange equation is a necessary condition for optimality in the context of the Basic Calculus of Variations Problem, where the boundary points are fixed but the curves are otherwise unconstrained. In this section we generalize that result to situations where equality constraints are imposed on the admissible curves. Before proceeding, the reader should find it helpful to review the material in Section 1.2.2 devoted to constrained optimality and the method of Lagrange multipliers for finite-dimensional problems. It is also useful to recall the general definitions of an extremum from Section 1.3.1 and of an admissible perturbation from Section 1.3.2; here the function space V is the same as at the beginning of Section 2.3 while the subset A is smaller because it reflects additional constraints (to be specified below). Finally, the explanation given at the beginning of Section 2.3 applies here as well: the conditions in this section are developed mainly with the 1-norm in mind, i.e., they are primarily designed to test for weak rather than strong minima.

2.5.1   Integral constraints

Suppose that we augment the Basic Calculus of Variations Problem with an additional constraint of the form

$$C(y) := \int_a^b M(x, y(x), y'(x))dx = C_0 \qquad (2.45)$$

where C stands for the "constraint" functional, M is a function from the same class as L, and C_0 is a given constant. In other words, the problem is to minimize the functional given by (2.9) over $\mathcal{C}^1$ curves $y(\cdot)$ satisfying the boundary conditions (2.8) and subject to the integral constraint (2.45). For simplicity, we are considering the case of only one constraint. We already saw examples of such constrained problems, namely, Dido's problem and the catenary problem.

Assume that a given curve y is an extremum. What follows is a *heuristic argument* motivated by our earlier derivation of the first-order necessary condition for constrained optimality in the finite-dimensional case (involving Lagrange multipliers). Let us consider perturbed curves of the familiar form

$$y + \alpha\eta.$$

To be admissible, the perturbation η must preserve the constraint (in addition to vanishing at the endpoints as before). In other words, we must have $C(y+\alpha\eta) = C_0$ for all α sufficiently close to 0. In terms of the first variation of C, this property is easily seen to imply that

$$\delta C|_y(\eta) = 0. \qquad (2.46)$$

Repeating the same calculation as in our original derivation of the Euler-Lagrange equation, we obtain from this that

$$\int_a^b \left(M_y(x, y(x), y'(x)) - \frac{d}{dx} M_{y'}(x, y(x), y'(x)) \right) \eta(x)dx = 0. \qquad (2.47)$$

Now our basic first-order necessary condition (1.36) indicates that for every η satisfying (2.47), we should have

$$\delta J|_y (\eta) = \int_a^b \left(L_y(x, y(x), y'(x)) - \frac{d}{dx} L_{y'}(x, y(x), y'(x)) \right) \eta(x) dx = 0.$$

This conclusion can be summarized as follows:

$$\int_a^b \left(L_y - \frac{d}{dx} L_{y'} \right) \eta(x) dx = 0 \quad \forall \eta \text{ such that } \int_a^b \left(M_y - \frac{d}{dx} M_{y'} \right) \eta(x) dx = 0.$$
(2.48)

The reader will note that (2.48) is quite similar to the condition (1.21) on page 12. It also has a similar consequence, namely, that there exists a constant λ^* (a *Lagrange multiplier*) such that

$$\left(L_y - \frac{d}{dx} L_{y'} \right) + \lambda^* \left(M_y - \frac{d}{dx} M_{y'} \right) = 0$$
(2.49)

for all $x \in [a, b]$. Rearranging terms, we see that this is equivalent to

$$(L + \lambda^* M)_y = \frac{d}{dx} (L + \lambda^* M)_{y'}$$

which amounts to saying that the Euler-Lagrange equation holds for the augmented Lagrangian $L + \lambda^* M$. In other words, y is an extremal of the augmented cost functional

$$(J + \lambda^* C)(y) = \int_a^b \left(L(x, y(x), y'(x)) + \lambda^* M(x, y(x), y'(x)) \right) dx.$$
(2.50)

A closer inspection of the above argument reveals, however, that we left a couple of gaps. First, we did not justify the step of passing from (2.48) to (2.49). In the finite-dimensional case, we had to make the corresponding step of passing from (1.21) to (1.22) which then gave (1.24); we would need to construct a similar reasoning here, treating the integrals in (2.48) as inner products of η with the functions in parentheses (inner products in $\mathcal{L}_2$). Second, there was actually a more serious logical flaw: the condition (2.46) is *necessary* for the perturbation η to preserve the constraint (2.45), but we do not know whether it is *sufficient*. Without this sufficiency, the validity of (2.48) is in serious doubt. In the finite-dimensional case, to reach (1.21) we used the fact that (1.20) was a necessary and sufficient condition for d to be a tangent vector; we did not, however, give a proof of the sufficiency part (which is not trivial).

It is also important to recall that in the finite-dimensional case studied in Section 1.2.2, the first-order necessary condition for constrained optimality in terms of Lagrange multipliers is valid only when an additional technical assumption holds, namely, the extremum must be a regular point of the constraint surface. This assumption is needed to rule out degenerate situations (see Exercise 1.2); in fact, it enables precisely the sufficiency part mentioned in the previous paragraph. It turns out that in the present case, a degenerate situation arises when the test curve y satisfies the constraint but all nearby curves violate it. This can happen if y is an extremal of the constraint functional C, i.e., satisfies the Euler-Lagrange equation for M. For example, consider the length constraint $C(y) := \int_0^1 \sqrt{1 + (y')^2} dx = 1$ together with the boundary conditions $y(0) = y(1) = 0$. Clearly, $y \equiv 0$ is the only admissible curve (it is the unique global minimum of the constraint functional), hence it automatically solves our constrained problem no matter what J is. The second integral in (2.48) is 0 for every η since y is an extremal of C. Thus if (2.48) were true, it would imply that y must be an extremal of J, but as we just explained this is not necessary. We see that if we hope for (2.48) to be a necessary condition for constrained optimality, we need to assume that y is not an extremal of C, so that there exist nearby curves at which C takes values both larger and smaller than C_0.

We can now *conjecture* the following **first-order necessary condition for constrained optimality**: *If $y(\cdot)$ is an extremum for the constrained problem and is not an extremal of the constraint functional C (i.e., does not satisfy the Euler-Lagrange equation for M), then it is an extremal of the augmented cost functional (2.50) for some $\lambda^* \in \mathbb{R}$.* We can also state this condition more succinctly, combining the nondegeneracy assumption and the conclusion into one statement: y must satisfy the Euler-Lagrange equation for $\lambda_0^* L + \lambda^* M$, where λ_0^* and λ^* are constants (not both 0). Indeed, this means that either $\lambda_0^* = 0$ and y is an extremal of C, or $\lambda_0^* \neq 0$ and y is an extremal of $J + (\lambda^*/\lambda_0^*)C$. The number λ_0^* is called the *abnormal multiplier* (it also has an analog in optimal control which will appear in Section 4.1).

It turns out that this conjecture is correct. However, rather than fixing the above faulty argument, it is easier to give an alternative proof by proceeding along the lines of the second proof in Section 1.2.2.

Exercise 2.9 *Write down a correct proof of the first-order necessary condition for constrained optimality by considering a two-parameter family of perturbed curves $y + \alpha_1 \eta_1 + \alpha_2 \eta_2$ and using the Inverse Function Theorem.* □

In the unconstrained case, as we noted earlier, the general solution of the second-order Euler-Lagrange differential equation depends on two arbitrary constants whose values are to be determined from the two boundary conditions. Here we have one additional parameter λ^* but also one additional constraint (2.45), so generically we still expect to obtain a unique extremal.

The generalization of the above necessary condition to problems with several constraints is straightforward: we need one Lagrange multiplier for each constraint (cf. Section 1.2.2). The multiple-degrees-of-freedom setting also presents no complications.

Similarly to the finite-dimensional case, Lagrange's original intuition was to replace constrained minimization of J with respect to y by unconstrained minimization of

$$\int_a^b L \, dx + \lambda \left(\int_a^b M \, dx - C_0 \right) \tag{2.51}$$

with respect to y and λ. For curves satisfying the constraint, the values of the two functionals coincide. However, for the same reasons as in the discussion on page 16, considering this augmented cost (which matches (2.50) except for an additive constant) does not lead to a rigorous justification of the necessary condition.

Equipped with the above necessary condition for the case of integral constraints as well as our previous experience with the Euler-Lagrange equation, we can now study Dido's isoperimetric problem and the catenary problem.

Exercise 2.10   *Show that optimal curves for Dido's problem are circular arcs and that optimal curves for the catenary problem satisfy (2.3).*   □

2.5.2   Non-integral constraints

We now suppose that instead of the integral constraint (2.45) we have an equality constraint which must hold pointwise:

$$M(x, y(x), y'(x)) = 0 \tag{2.52}$$

for all $x \in [a, b]$. A vector-valued function M can be used to describe multiple constraints, but here we assume for simplicity that there is only one constraint.

Let y be a test curve. It turns out that the first-order necessary condition for optimality in this case is similar to that for integral constraints, but the Lagrange multiplier is now a function of x. In other words, the Euler-Lagrange equation must hold for the augmented Lagrangian

$$L + \lambda^*(x) M$$

where $\lambda^* : [a, b] \to \mathbb{R}$ is some function. As in Section 2.5.1, we need a technical assumption to rule out degenerate cases. Here we need to assume that there are at least two degrees of freedom and that everywhere along the curve we have $M_{y'} \neq 0$ or, if y' does not appear in (2.52), $M_y \neq 0$. (This permits us, via the Implicit Function Theorem, to locally solve for one dependent variable component in (2.52) in terms of the others and to guarantee the existence of other curves near y satisfying the constraint.)

We will not give a proof of the above result, and instead limit ourselves to an intuitive explanation. The integral constraint (2.45) is global, in the sense that it applies to the entire curve. In contrast, the non-integral constraint (2.52) is local, i.e., applies to each point on the curve. Locally around each point, there is no essential difference between the two. This suggests that for each x there should exist a Lagrange multiplier, and these can be pieced together to give the desired function $\lambda^* = \lambda^*(x)$. Another way to see the correspondence between the two problems is to follow Lagrange's idea of considering an augmented cost functional which coincides with the original one for curves satisfying the constraint. For integral constraints this was accomplished by (2.51), while here we can consider the minimization of

$$\int_a^b L\,dx + \int_a^b \lambda(x)M\,dx \tag{2.53}$$

over y and λ. Note that λ no longer needs to be constant, since M is identically 0. (The role of integration in the second term is simply to obtain a cost for the whole curve.)

What if we want to solve the constraint (2.52) for y' as a function of x and y? In general, we have fewer constraints than the dimension of y', i.e., the system of equations is under-determined. This means that we will have some free parameters; denoting by u the vector of these parameters, we can write the solution in the form

$$y' = f(x, y, u).$$

The reader hopefully recognizes this as a control system, in the disguise of a somewhat unfamiliar notation compared to (1.1). For example, if $n = 2$ (two degrees of freedom) and $M(x, y, y') = y_1' - y_2$, then we have $y_1' = y_2$ and $y_2' = u$ (free). Incidentally, the case of no constraints (our Basic Calculus of Variations Problem) corresponds to $y' = u$. We will turn our attention to control problems later in the book, and the present lack of depth in our treatment of non-integral constraints will be amply compensated for. In particular, the "distributed" Lagrange multiplier $\lambda^*(\cdot)$ will correspond to the adjoint vector, or costate, in the maximum principle.

HOLONOMIC CONSTRAINTS

Consider the special case when the constraint function M does not depend on y', so that the constraints take the form

$$M(x, y(x)) = 0. \tag{2.54}$$

Alternatively, we might be able to integrate the constraints (2.52) to bring them to this form. We then say that the constraints are *holonomic*. The equation (2.54) gives us a constraint surface in the (x, y)-space, and we have two options for studying our constrained optimization problem. The first one is to use the previous necessary condition involving a Lagrange multiplier function. The second option is to find fewer independent variables that parameterize the constraint surface, and reformulate the problem in terms of these variables. The problem then becomes an unconstrained one, and can be studied via the usual Euler-Lagrange equation. Sometimes this latter approach turns out to be more effective.

Example 2.5 *Consider a simple planar pendulum depicted in Figure 2.11. The center of coordinates is attached to the pivot point. The length of the pendulum is ℓ and the mass at the tip is m. We want to derive trajectories of motion for this system, using the principle of least action. Since this is a mechanical example, we use the time t as the independent variable, treating x and y as the dependent variables (see Section 2.4.3).*

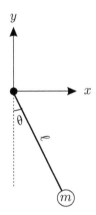

Figure 2.11: The pendulum

The constraint is $M(x, y) := x^2 + y^2 - \ell^2 = 0$, which is holonomic. In polar coordinates (r, θ), this means that $r \equiv \ell$ while the angle θ is free. Let us follow the second approach described above and express everything in terms of θ as a single degree of freedom, writing $\dot{\theta}$ for $d\theta/dt$. The kinetic energy is $T = \frac{1}{2}m\ell^2\dot{\theta}^2$ (this follows from the formula (2.44) at the end of Section 2.4.3). The potential energy is $U = mg\ell(1 - \cos\theta)$, up to an additive constant (the present choice gives $U = 0$ at the downward equilibrium). Thus the Lagrangian is $L = T - U = \frac{1}{2}m\ell^2\dot{\theta}^2 + mg\ell(\cos\theta - 1)$ and the Euler-Lagrange equation (2.43) gives

$$\ddot{\theta} = -\frac{g}{\ell}\sin\theta \qquad (2.55)$$

which is the familiar pendulum equation.   □

Exercise 2.11 *Study the above example directly in the (x, y)-coordinates, with the help of the necessary condition for constrained optimality involving a Lagrange multiplier function. Go as far as you can in deriving the equations of motion. Are you able to reproduce (2.55) using this method?* ☐

In control theory, one is often more interested in the opposite situation where the constraints are nonholonomic, i.e., cannot be integrated. In this case, two arbitrary points in the (x, y)-space can be connected by a path satisfying the constraints, and there is no lower-dimensional constraint surface.

2.6 SECOND-ORDER CONDITIONS

In Section 2.3 we used the first variation to obtain the first-order necessary condition for optimality expressed by the Euler-Lagrange equation. In this section we will work with the second variation and derive first a necessary condition and then a sufficient condition for optimality. The setting is that of the Basic Calculus of Variations Problem and weak minima (cf. the discussion at the beginning of Section 2.3).

Recall from Section 1.3.3 the second-order expansion

$$J(y + \alpha\eta) = J(y) + \delta J\big|_y (\eta)\alpha + \delta^2 J\big|_y (\eta)\alpha^2 + o(\alpha^2) \tag{2.56}$$

which defines the quadratic form $\delta^2 J\big|_y$ called the second variation. The basic second-order necessary condition (1.39) says that if a curve y is a local minimum of J, then for all admissible perturbations η we must have $\delta^2 J\big|_y (\eta) \geq 0$. In the present context, the function space V and its subset A are as at the beginning of Section 2.3, and so admissible perturbations are $\mathcal{C}^1$ functions satisfying (2.11). Using the fact that the functional J takes the form (2.9), we will derive in Section 2.6.1 a more explicit second-order necessary condition for this situation.

We also discussed in Section 1.3.3 that in order to develop a second-order *sufficient* condition for a local minimum, we need to assume the first-order necessary condition, strengthen the second-order necessary condition to $\delta^2 J\big|_y (\eta) > 0$, and then try to prove that the second-order term dominates the higher-order term in the expansion (2.56). For the variational problem at hand, we will be able to carry out this program in Section 2.6.2.

Of course, second-order conditions for a local *maximum* are easily obtained by reversing the inequalities (or, equivalently, by replacing J with $-J$). For this reason, we confine our attention to minima.

2.6.1   Legendre's necessary condition for a weak minimum

Let us compute $\delta^2 J\big|_y$ for a given test curve y. Since third-order partial derivatives of L will appear, we assume that $L \in \mathcal{C}^3$. We work with the single-degree-of-freedom case for now. The left-hand side of (2.56) is

$$J(y + \alpha\eta) = \int_a^b L(x, y(x) + \alpha\eta(x), y'(x) + \alpha\eta'(x))dx.$$

We need to write down its second-order Taylor expansion with respect to α. We do this by expanding the function inside the integral with respect to α (using the chain rule) and separating the terms of different orders in α:

$$J(y + \alpha\eta) = \int_a^b L(x, y(x), y'(x))dx + \alpha \int_a^b \big(L_y(x, y(x), y'(x))\eta(x)$$
$$+ L_{y'}(x, y(x), y'(x))\eta'(x)\big)dx + \frac{\alpha^2}{2} \int_a^b \big(L_{yy}(x, y(x), y'(x))(\eta(x))^2$$
$$+ 2L_{yy'}(x, y(x), y'(x))\eta(x)\eta'(x) + L_{y'y'}(x, y(x), y'(x))(\eta'(x))^2\big)dx$$
$$+ o(\alpha^2). \tag{2.57}$$

Matching this expression with (2.56) term by term, we deduce that the second variation is given by

$$\delta^2 J\big|_y(\eta) = \frac{1}{2} \int_a^b \big(L_{yy}\eta^2 + 2L_{yy'}\eta\eta' + L_{y'y'}(\eta')^2\big)dx$$

where the integrand is evaluated along $(x, y(x), y'(x))$. This is indeed a quadratic form as defined in Section 1.3.3. Note that it explicitly depends on η' as well as on η. In contrast with the first variation (analyzed in detail in Section 2.3.1), the dependence of the second variation on η' is essential and cannot be eliminated. We can, however, simplify the expression for $\delta^2 J\big|_y(\eta)$ by eliminating the "mixed" term containing the product $\eta\eta'$. We do this by using—as we did in our earlier derivation of the Euler-Lagrange equation—the method of integration by parts:

$$\int_a^b 2L_{yy'}\eta\eta'dx = \int_a^b L_{yy'}\frac{d}{dx}(\eta^2)dx = L_{yy'}\eta^2\big|_a^b - \int_a^b \frac{d}{dx}(L_{yy'})\eta^2dx.$$

The first, non-integral term on the right-hand side vanishes due to the boundary conditions (2.11). Therefore, the second variation can be written as

$$\delta^2 J\big|_y(\eta) = \int_a^b \big(P(x)(\eta'(x))^2 + Q(x)(\eta(x))^2\big)\,dx \tag{2.58}$$

where

$$P(x) := \frac{1}{2}L_{y'y'}(x, y(x), y'(x)),$$
$$Q(x) := \frac{1}{2}\Big(L_{yy}(x, y(x), y'(x)) - \frac{d}{dx}L_{yy'}(x, y(x), y'(x))\Big). \tag{2.59}$$

Note that P is continuous, and Q is also continuous at least when $y \in \mathcal{C}^2$.

When we come to the issue of sufficiency in the next subsection, we will also need a more precise characterization of the higher-order term labeled as $o(\alpha^2)$ in the expansion (2.57). The next exercise invites the reader to go back to the derivation of (2.57) and analyze this term in more detail.

Exercise 2.12   *Use Taylor's theorem with remainder (see, e.g., [Rud76, Theorem 5.15]) to show that the $o(\alpha^2)$ term in (2.57) can be written in the form*

$$o(\alpha^2) = \alpha^2 \int_a^b \left(\bar{P}(x, \eta(x), \eta'(x), \alpha)(\eta'(x))^2 + \bar{Q}(x, \eta(x), \eta'(x), \alpha)(\eta(x))^2 \right) dx$$

(2.60)

where $\bar{P}, \bar{Q} \to 0$ as $\alpha \to 0$. Moreover, show that this convergence is uniform over η with respect to the 1-norm, in the sense that for every $\gamma > 0$ there exists an $\varepsilon > 0$ such that for $|\alpha| < \varepsilon$ we have $|\bar{P}(x, \eta(x), \eta'(x), \alpha)| < \gamma$ and $|\bar{Q}(x, \eta(x), \eta'(x), \alpha)| < \gamma$ for all η with $\|\eta\|_1 = 1$ and all $x \in [a, b]$, but the same statement with respect to the 0-norm is in general false.   ☐

We know that if y is a minimum, then for all $\mathcal{C}^1$ perturbations η vanishing at the endpoints the quantity (2.58) must be nonnegative:

$$\int_a^b \left(P(x)(\eta'(x))^2 + Q(x)(\eta(x))^2 \right) dx \geq 0.$$

(2.61)

We would like to restate this condition in terms of P and Q only—which are defined directly from L and y via (2.59)—so that we would not need to check it for all η. (Recall that we followed a similar route earlier when passing from the condition (2.16) to the Euler-Lagrange equation via Lemma 2.1.)

What, if anything, does the inequality (2.61) imply about P and Q? Does it force at least one of these two functions, or perhaps both, to be nonnegative on $[a, b]$? The two terms inside the integral in (2.61) are of course not independent because η' and η are related. So, we should try to see if maybe one of them dominates the other. More specifically, can it happen that η' is large (in magnitude) while η is small, or the other way around?

To answer these questions, consider a family of perturbations η_ε parameterized by small $\varepsilon > 0$, depicted in Figure 2.12. The function η_ε equals 0 everywhere outside some interval $[c, d] \subset [a, b]$, and inside this interval it equals 1 except near the endpoints where it rapidly goes up to 1 and back down to 0. This rapid transfer is accomplished by the derivative η_ε' having a short pulse of width approximately ε and height approximately $1/\varepsilon$ right after c, and a similar negative pulse right before d. Here ε is small compared to $d - c$. We base the subsequent argument on this graphical description, but it is not difficult to specify a formula for η_ε and use it to verify the claims that follow; see, e.g., [GF63, p. 103] for a similar construction.

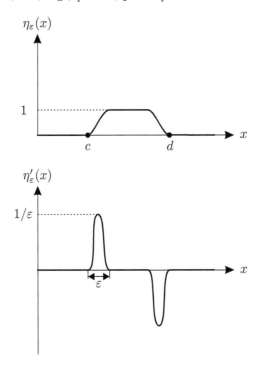

Figure 2.12: The graphs of η_ε and its derivative

We can see that

$$\left| \int_a^b Q(x)(\eta_\varepsilon(x))^2 dx \right| \le \int_c^d |Q(x)| dx$$

and this bound is uniform over ε. On the other hand, for nonzero P the integral $\int_a^b P(x)(\eta'_\varepsilon(x))^2 dx$ does not stay bounded as $\varepsilon \to 0$, because it is of order $1/\varepsilon$. In particular, let us see this more clearly for the case when P is negative on $[c, d]$, so that for some $\delta > 0$ we have $P(x) \le -\delta$ for all $x \in [c, d]$. Assume that there is an interval inside $[c, c + \varepsilon]$ of length at least $\varepsilon/2$ on which η'_ε is no smaller than $1/(2\varepsilon)$; this property is completely consistent with our earlier description of η_ε and η'_ε. We then have

$$\int_a^b P(x)(\eta'_\varepsilon(x))^2 dx \le \int_c^{c+\varepsilon} P(x)(\eta'_\varepsilon(x))^2 dx \le -\delta \frac{1}{4\varepsilon^2} \frac{\varepsilon}{2} = -\frac{\delta}{8\varepsilon}.$$

As $\varepsilon \to 0$, the above expression tends to $-\infty$, dominating the bounded Q-dependent term. It follows that the inequality (2.61) cannot hold for all η if P is negative on some subinterval $[c, d]$. But this means that for y to be a minimum, P must be nonnegative everywhere on $[a, b]$. Indeed, if $P(\bar{x}) < 0$ for some $\bar{x} \in [a, b]$, then by continuity of P we can find a subinterval $[c, d]$

containing $\bar{x}$ on which P is negative, and the above construction can be applied.[3]

Recalling the definition of P in (2.59), we arrive at our **second-order necessary condition for optimality**: *For all $x \in [a, b]$ we must have*

$$\boxed{L_{y'y'}(x, y(x), y'(x)) \geq 0} \tag{2.62}$$

This condition is known as **Legendre's condition**, as it was obtained by Legendre in 1786. Note that it places no restrictions on the sign of Q; intuitively speaking, the Q-dependent term in the second variation is dominated by the P-dependent term, hence Q can in principle be negative along the optimal curve (this point will become clearer in the next subsection). For multiple degrees of freedom, the proof takes a bit more work but the statement of Legendre's condition is virtually unchanged: $L_{y'y'}(x, y(x), y'(x))$, which becomes a symmetric matrix, must be positive semidefinite for all x, i.e., (2.62) must hold in the matrix sense along the optimal curve.

As a brief digression, let us recall our definition (2.29) of the Hamiltonian:

$$H(x, y, y', p) = p \cdot y' - L(x, y, y') \tag{2.63}$$

where $p = L_{y'}(x, y, y')$. We already noted in Section 2.4.1 that when H is viewed as a function of y' with the other arguments evaluated along an optimal curve y, it should have a stationary point at $y' = y'(x)$, the velocity of y at x. More, precisely, the function H^* defined in (2.32) has a stationary point at $z = y'(x)$, which is a consequence of (2.33). Legendre's condition tells us that, in addition, $H_{y'y'} = -L_{y'y'} \leq 0$ along an optimal curve, which we can rewrite in terms of H^* as

$$\frac{d^2 H^*}{dz^2}(y'(x)) = -L_{y'y'}(x, y(x), y'(x)) \leq 0.$$

Thus, if the above stationary point is an extremum, then it is necessarily a maximum. This interpretation of necessary conditions for optimality moves us one step closer to the maximum principle. The basic idea behind our derivation of Legendre's condition will reappear in Section 3.4 in the context of optimal control, but eventually (in Chapter 4) we will obtain a stronger result using more advanced techniques.

2.6.2   Sufficient condition for a weak minimum

We are now interested in obtaining a second-order sufficient condition for proving optimality of a given test curve y. Looking again at the expansion (2.56) and recalling our earlier discussions, we know that we want to

---

[3]Note that while $\|\eta_\varepsilon\|_1$ is large for small ε, once ε is fixed we have $\|\alpha\eta\|_1 \to 0$ as $\alpha \to 0$, so the perturbed curves are still close to y in the sense of the 1-norm.

have $\delta^2 J|_y(\eta) > 0$ for all admissible perturbations, which means having a strict inequality in (2.61). In addition, we need some uniformity to be able to dominate the $o(\alpha^2)$ term. Since we saw that the P-dependent term inside the integral in (2.61) is the dominant term in the second variation, it is natural to conjecture—as Legendre did—that having $P(x) > 0$ for all $x \in [a, b]$ should be sufficient for the second variation to be positive definite. Legendre tried to prove this implication using the following clever approach. For every differentiable function $w = w(x)$ we have

$$0 = w\eta^2\big|_a^b = \int_a^b \frac{d}{dx}(w\eta^2)dx = \int_a^b (w'\eta^2 + 2w\eta\eta')dx$$

where the first equality follows from the constraint $\eta(a) = \eta(b) = 0$. This lets us rewrite the second variation as

$$\int_a^b \big(P(x)(\eta'(x))^2 + Q(x)(\eta(x))^2\big)dx$$
$$= \int_a^b \big(P(x)(\eta'(x))^2 + 2w(x)\eta(x)\eta'(x) + (Q(x) + w'(x))(\eta(x))^2\big)dx.$$

Now, the idea is to find a function w that makes the integrand on the right-hand side into a perfect square. Clearly, such a w needs to satisfy

$$P(Q + w') = w^2. \tag{2.64}$$

This is a quadratic differential equation, of *Riccati* type, for the unknown function w.

Let us suppose that we found a function w satisfying (2.64). Then our second variation can be written as

$$\int_a^b \left(\sqrt{P(x)}\eta'(x) + \frac{w(x)}{\sqrt{P(x)}}\eta(x)\right)^2 dx = \int_a^b P(x)\left(\eta'(x) + \frac{w(x)}{P(x)}\eta(x)\right)^2 dx \tag{2.65}$$

(the division by P is permissible since we are operating under the assumption that $P > 0$). It is obvious that the right-hand side of (2.65) is nonnegative, but we claim that it is actually positive for every admissible perturbation η that is not identically 0. Indeed, if the integral is 0, then $\eta' + \frac{w}{P}\eta \equiv 0$. We also know that $\eta(a) = 0$. But there is only one solution $\eta : [a, b] \to \mathbb{R}$ of the first-order differential equation $\eta' + \frac{w}{P}\eta = 0$ with the zero initial condition, and this solution is $\eta \equiv 0$. So, it seems that we have $\delta^2 J|_y(\eta) > 0$ for all $\eta \not\equiv 0$. At this point we challenge the reader to see a gap in the above argument.

The problem with the foregoing reasoning is that the Riccati differential equation (2.64) may have a *finite escape time*, i.e., the solution w may not exist on the whole interval $[a, b]$. For example, if $P \equiv 1$ and $Q \equiv -1$ then (2.64) becomes $w' = w^2 + 1$. Its solution $w(x) = \tan(x - c)$, where the constant c depends on the choice of the initial condition, blows up when $x - c$ is an odd integer multiple of $\pi/2$. This means that w will not exist on all of $[a, b]$ for any choice of $w(a)$ if $b - a \geq \pi$.

We see that a sufficient condition for optimality should involve, in addition to an inequality like $L_{y'y'} > 0$ holding pointwise along the curve, some "global" considerations applied to the entire curve. In fact, this becomes intuitively clear if we observe that a concatenation of optimal curves is not necessarily optimal. For example, consider the two great-circle arcs on a sphere shown in Figure 2.13. Each arc minimizes the distance between its endpoints, but this statement is no longer true for their concatenation—even when compared with nearby curves. At the same time, the concatenated arc would still satisfy any pointwise condition fulfilled by the two pieces.

Figure 2.13: Concatenation of shortest-distance curves on a sphere is not shortest-distance

Returning to our analysis, we need to ensure the existence of a solution for the differential equation (2.64) on the whole interval $[a, b]$. This issue, which escaped Legendre's attention, was pointed out by Lagrange in 1797. However, it was only in 1837, after 50 years had passed since Legendre's investigation, that Jacobi closed the gap by providing a missing ingredient which we now describe. The first step is to reduce the quadratic first-order differential equation (2.64) to another differential equation, linear but of second order, by making the substitution

$$w(x) = -\frac{Pv'(x)}{v(x)} \tag{2.66}$$

where v is a new (unknown) function, twice differentiable and not equal to 0 anywhere. Rewriting (2.64) in terms of v, we obtain

$$P\left(Q - \frac{\frac{d}{dx}(Pv')v - P(v')^2}{v^2}\right) = \frac{P^2(v')^2}{v^2}.$$

Multiplying both sides of this equation by v (which is nonzero), dividing by P (which is positive), and canceling terms, we can bring it to the form

$$Qv = \frac{d}{dx}(Pv').$$ (2.67)

This is the so-called *accessory*, or *Jacobi*, equation. We will be done if we can find a solution v of the accessory equation (2.67) that does not vanish anywhere on $[a, b]$, because then we can obtain a desired solution w to the original equation (2.64) via the formula (2.66).

Since (2.67) is a second-order differential equation, the initial data at $x = a$ needed to uniquely specify a solution consists of $v(a)$ and $v'(a)$. In addition, note that if v is a solution of (2.67) then λv is also a solution for every constant λ. By adjusting λ appropriately, we can thus assume with no loss of generality that $v'(a) = 1$ (since we are not interested in v being identically 0). Among such solutions, let us consider the one that starts at 0, i.e., set $v(a) = 0$. A point $c > a$ is said to be *conjugate* to a if this solution v hits 0 again at c, i.e., $v(c) = v(a) = 0$ (see Figure 2.14). It is clear that conjugate points are completely determined by P and Q, which in turn depend, through (2.59), only on the test curve y and the Lagrangian L in the original variational problem.

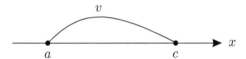

Figure 2.14: A conjugate point

Conjugate points have a number of interesting properties and interpretations, and their theory is outside the scope of this book. We do mention the following interesting fact, which involves a concept that we will see again later when proving the maximum principle. If we consider two neighboring extremals (solutions of the Euler-Lagrange equation) starting from the same point at $x = a$, and if c is a point conjugate to a, then at $x = c$ the distance between these two extremals becomes small (an infinitesimal of higher order) relative to the distance between the two extremals as well as between their derivatives over $[a, b]$. As their distance over $[a, b]$ approaches 0, the two extremals actually intersect at a point whose x-coordinate approaches c. The reason behind this phenomenon is that the Jacobi equation is, approximately, the differential equation satisfied by the difference between two neighboring extremals; the next exercise makes this statement precise.

Exercise 2.13 *Suppose that y and $y + v$ are two neighboring extremals of the functional (2.9). Show that then v must satisfy*

$$Qv - \frac{d}{dx}(Pv') = o(\|v\|)$$ (2.68)

where P and Q are as in (2.59) and $\|\cdot\|$ is a suitable norm (specify which one). $\qquad\square$

We see from (2.68) that v, which is the difference between the two extremals, satisfies the Jacobi equation (2.67) modulo terms of higher order. A linear differential equation that describes, within terms of higher order, the propagation of the difference between two nearby solutions of a given differential equation is called the *variational equation* (corresponding to the given differential equation). In this sense, the Jacobi equation is the variational equation for the Euler-Lagrange equation. This property can be shown to imply the claims we made before the exercise. Intuitively speaking, a conjugate point is where different neighboring extremals starting from the same point meet again (approximately). If we revisit the example of shortest-distance curves on a sphere, we see that conjugate points correspond to diametrically opposite points: all extremals (which are great-circle arcs) with a given initial point intersect after completing half a circle. We will encounter the concept of a variational equation again in Section 4.2.4.

Now, suppose that the interval $[a, b]$ contains no points conjugate to a. Let us see how this may help us in our task of finding a solution v of the Jacobi equation (2.67) that does not equal 0 anywhere on $[a, b]$. The absence of conjugate points means, by definition, that the solution with the initial data $v(a) = 0$ and $v'(a) = 1$ never returns to 0 on $[a, b]$. This is not yet a desired solution because we cannot have $v(a) = 0$. What we can do, however, is make $v(a)$ very small but positive. Using the property of continuity with respect to initial conditions for solutions of differential equations, it is possible to show that such a solution will remain positive everywhere on $[a, b]$.

In view of our earlier discussion, we conclude that the second variation $\delta^2 J\big|_y$ is positive definite (on the space of admissible perturbations) if $P(x) > 0$ for all $x \in [a, b]$ and there are no points conjugate to a on $[a, b]$. We remark in passing that the absence of points conjugate to a on $[a, b]$ is also a necessary condition for $\delta^2 J\big|_y$ to be positive definite, and if $\delta^2 J\big|_y$ is positive semidefinite then no interior point of $[a, b]$ can be conjugate to a. We are now ready to state the following **second-order sufficient condition for optimality**: *An extremal $y(\cdot)$ is a strict minimum if $L_{y'y'}(x, y(x), y'(x)) > 0$ for all $x \in [a, b]$ and the interval $[a, b]$ contains no points conjugate to a.*

Note that we do not yet have a proof of this result. Referring to the second-order expansion (2.56), we know that under the conditions just listed $\delta J\big|_y(\eta) = 0$ (since y is an extremal) and, as we just saw, $\delta^2 J\big|_y(\eta)$ given by (2.58) is positive (unless $\eta \equiv 0$). However, we still need to show that $\delta^2 J\big|_y(\eta)\alpha^2$ dominates the higher-order term $o(\alpha^2)$ which has the properties established in Exercise 2.12. Since $P(x) = \frac{1}{2}L_{y'y'}(x, y(x), y'(x)) > 0$ on

$[a, b]$, we can pick a small enough $\delta > 0$ such that $P(x) > \delta$ for all $x \in [a, b]$. Consider the integral

$$\int_a^b \left((P(x) - \delta)(\eta'(x))^2 + Q(x)(\eta(x))^2 \right) dx. \tag{2.69}$$

Reducing δ further towards 0 if necessary, we can ensure that no points conjugate to a on $[a, b]$ are introduced as we pass from P to $P - \delta$ (thanks to continuity of solutions of the accessory equation with respect to parameter variations). This guarantees that the functional (2.69) is still positive definite, hence

$$\int_a^b \left(P(x)(\eta'(x))^2 + Q(x)(\eta(x))^2 \right) dx > \delta \int_a^b (\eta'(x))^2 dx \tag{2.70}$$

for all admissible perturbations (not identically equal to 0).

In light of our earlier derivation of Legendre's condition, we know that the term depending on $(\eta')^2$ is in some sense the dominant term in (2.60), and the inequality (2.70) indicates that we are in good shape. Formally, we can handle the other, η^2-dependent term in (2.60) as follows. Use the Cauchy-Schwarz inequality with respect to the $\mathcal{L}_2$ norm[4] to write

$$\eta^2(x) = \left(\int_a^x 1 \cdot \eta'(z) dz \right)^2 \leq (x - a) \int_a^x (\eta'(z))^2 dz \leq (x - a) \int_a^b (\eta'(z))^2 dz.$$

From this, we have

$$\int_a^b \eta^2(x) dx \leq \int_a^b (x - a) dx \int_a^b (\eta'(z))^2 dz = \frac{(b - a)^2}{2} \int_a^b (\eta'(x))^2 dx. \tag{2.71}$$

Now, Exercise 2.12 tells us that the term $o(\alpha^2)$ in (2.56) takes the form (2.60) where for α close enough to 0 both $|\bar{P}|$ and $|\bar{Q}(b - a)^2/2|$ are smaller than $\delta/2$ for all $x \in [a, b]$ and all η with $\|\eta\|_1 = 1$. Combined with (2.58), (2.70), and (2.71) this implies $J(y + \alpha\eta) > J(y)$ for these values of α (except of course $\alpha = 0$), proving that y is a (strict) weak minimum.

The above sufficient condition is not as constructive and practical as the first-order and second-order necessary conditions, because to apply it one needs to study conjugate points. The simpler necessary conditions can be exploited first, to see if they help narrow down candidates for an optimal solution. It should be observed, though, that the existence of conjugate points can be ruled out if the interval $[a, b]$ is taken to be sufficiently small.

Exercise 2.14 *Justify the term "principle of least action" by showing that extremals of the action integral considered in Section 2.4.3 are automatically its minima on sufficiently small time intervals.* □

---

[4]I.e., $\left(\int f(x) g(x) dx \right)^2 \leq \int f^2(x) dx \cdot \int g^2(x) dx.$

As for the multiple-degrees-of-freedom setting, let us make the simplifying assumption that $L_{yy'}$ is a symmetric matrix (i.e., $L_{y_i y_j'} = L_{y_j y_i'}$ for all $i, j \in \{1, \ldots, n\}$). Then it is not difficult to show, following steps similar to those that led us to (2.58), that the second variation $\delta^2 J\big|_y$ is given by the formula

$$\delta^2 J\big|_y (\eta) = \int_a^b \left((\eta')^T (x) P(x) \eta'(x) + \eta^T(x) Q(x) \eta(x) \right) dx$$

where $P(x)$ and $Q(x)$ are symmetric matrices still defined by (2.59). In place of w introduced at the beginning of this subsection we need to consider a symmetric matrix W, and a suitable modification of our earlier square completion argument yields the Riccati matrix differential equation

$$Q + W' = W P^{-1} W$$

(W' denotes the derivative of W, not the transpose). This quadratic differential equation is reduced to the second-order linear matrix differential equation $QV = \frac{d}{dx}(PV')$ by the substitution $W = -PV'V^{-1}$, where V is a matrix. Conjugate points are defined in terms of V becoming singular. Generalizing the previous results by following this route is straightforward. Riccati matrix differential equations and their solutions play a central role in the linear quadratic regulator problem, which we will study in detail in Chapter 6.

2.7   NOTES AND REFERENCES FOR CHAPTER 2

Formulations and solutions of the problems of Dido, catenary, and brachistochrone, as well as related historical remarks, are given in [GF63, You80, Mac05] and many other sources. For an enlightening discussion of light reflection and refraction, see [FLS63, Chapter I-26], where there is also an amusing (although perhaps not entirely politically correct) alternative description of refraction in terms of choosing the fastest path from the beach to the water to save a drowning girl. A comprehensive, insightful, and mathematically accurate account of the historical development of calculus of variations is given in [Gol80]; this book traces the roots of the subject to Fermat's principle of least time, which allowed the use of calculus for analyzing light refraction and later inspired Johann Bernoulli's solution of the brachistochrone problem. For an in-depth treatment of the brachistochrone problem we also recommend the paper [SW97]; it is explained there that this problem effectively marked the birth of the field of optimal control, because it started steady research activity on time-optimal and related variational problems still studied today in optimal control theory. Regarding the catenary, it is interesting to mention that the inverted catenary shape has been used for building arches from ancient times to present day (notable examples include several buildings designed by Gaudi in Catalonia and the Gateway Arch in St. Louis, Missouri).

Sections 2.2 and 2.3 follow Chapter 1 of [GF63], the text on which most of the present chapter is based; see also [Mac05]. Exercise 2.2 is borrowed from [Jur96, p. 341]. Example 2.3 is treated in [Vin00, p. 18], where it is followed by a discussion of existence of optimal solutions. Differentiability assumptions under which the Euler-Lagrange equation is valid are discussed, in addition to [GF63] and [Mac05], in [Sus00, Handout 2]. Invariance of the Euler-Lagrange equation under changes of coordinates is demonstrated in [GF63] for a single degree of freedom and in [Sus00, Handout 2] for multiple degrees of freedom. The treatment of variable-endpoint problems in [GF63] includes the case of both endpoints lying on given vertical lines, as well as a variable-terminal-point version of the brachistochrone problem. A more general study leading to transversality conditions can be found in [GF63, Chapter 3] (although it relies on the general formula for the variation of a functional which we do not give) and in [SW77, Chapter 3].

The material of Section 2.4 is covered in [GF63, Chapter 4]; the Hamiltonian and the canonical variables also appear in [GF63, Chapter 3] in the general formula for the variation of a functional. Other sources of relevant information include [Arn89], [Mac05], and [Sus00, Handout 3]. Section 14 of [Arn89] mentions several applications of the Legendre transformation, including an elegant derivation of Young's inequality. Convexity of the Legendre transform (which is also called the conjugate function) is used in dual optimization methods; see [BV04, Sections 3.3 and 5.1]. The symmetric way of writing the Legendre transformation via the formula (2.36) is prompted by the presentation in [YZ99, pp. 220–221]. All these references also cover the Legendre transformation for functions of several variables. For an insightful discussion of how the Hamiltonian should be interpreted and why the maximum principle was not discovered much earlier, see [Sus00, Handout 3] and [SW97]. A nice exposition of the principle of least action can be found in [FLS63, Chapter II-19], and the reader intrigued by our brief remark about Einstein's theory of gravitation is advised to check Chapter II-42 of the same book. Conservation laws are derived in [GF63, Chapter 4] with the help of Noether's theorem, which is another application of the general formula for the variation of a functional. Conservation of angular momentum is also discussed in detail in [Arn89] (see in particular Example 2 in Section 13).

Section 2.5 is based on [GF63, Section 12] and [Mac05, Chapter 5], where additional details (such as the treatment of several integral constraints and multiple degrees of freedom, as well as a derivation of the necessary condition for the case of holonomic non-integral constraints) can be found. The book [You80] examines Lagrange's naive argument in detail and criticizes "its reappearance, every so often, in so-called accounts and introductions that claim to present the calculus of variations to engineers and other supposedly uncritical persons" (see the preamble to Volume II). The pendulum example is discussed in [Mac05, pp. 37 and 83]. For more information on control systems with nonholonomic constraints, including optimal control problems, the reader can consult [Blo03] and the references therein.

Section 2.6 is largely subsumed by Chapter 5 of [GF63]. Among additional topics covered there are a detailed study of conjugate points, a derivation of Legendre's condition for multiple degrees of freedom, and a connection with Sylvester's positive definiteness criterion for quadratic forms. It is possible to prove Legendre's condition for multiple degrees of freedom differently from how it is done in [GF63], without integrating by parts and without assuming that the matrix $L_{yy'}$ is symmetric; namely, one can perturb y along directions of eigenvectors of $L_{y'y'}$ (at an arbitrary fixed point x) and then invoke the scalar result, as in [LL50, pp. 94–95]. Our reasoning that the absence of conjugate points leads to the existence of a nonzero solution of the Jacobi equation is close to the argument given in [Mac05, Section 9.4], where the reader can find some missing details.

Chapter Three

From Calculus of Variations to Optimal Control

3.1 NECESSARY CONDITIONS FOR STRONG EXTREMA

As explained at the beginning of Section 2.3 and elsewhere in Chapter 2, the methods and results discussed in that chapter apply primarily to weak minima over C^1 curves. On the other hand, we illustrated in Section 2.2.1 that ultimately—especially in the context of optimal control—we are more interested in studying stronger notions of local optimality over less regular curves. In the next chapter we will realize this objective with the help of the maximum principle. The present chapter serves as a bridge between calculus of variations and the maximum principle. In this section, we present two results on strong minima over piecewise C^1 curves for the Basic Calculus of Variations Problem. In deriving these results we will depart, for the first time, from the familiar family of perturbed curves (2.10). The maximum principle will require a somewhat more advanced technical machinery than what we have seen so far, and we will now start "warming up" for it. An added benefit is that we will be able to maintain continuity in tracing the historical development of the subject. As we will see, from the maximum principle we will be able to recover the results given here, and more.

3.1.1 Weierstrass-Erdmann corner conditions

Recall from Section 2.2.1 that a piecewise C^1 curve y on $[a, b]$ is C^1 everywhere except possibly at a finite number of points where it is continuous but its derivative y' is discontinuous. Such points of discontinuity of y' are known as *corner points*. A corner point $c \in [a, b]$ is characterized by the property that the left-hand derivative $y'(c^-) := \lim_{x \nearrow c} y'(x)$ and the right-hand derivative $y'(c^+) := \lim_{x \searrow c} y'(x)$ both exist but have different values. For example, if a hanging chain (catenary) is suspended too close to the ground, then it will not look as in Figure 2.3 on page 29 but will instead touch the ground and have two corner points; see Figure 3.1. Below is another example in which a corner point arises.

Example 3.1 *Consider the problem of minimizing the functional $J(y) = \int_{-1}^{1} y^2(x)(y'(x)-1)^2 dx$ subject to the boundary conditions $y(-1) = 0$, $y(1) =$*

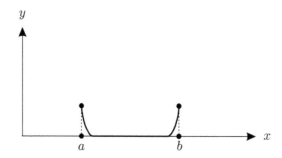

Figure 3.1: An extremal with two corners for the catenary problem

1. *It is clear that $J(y) \geq 0$ for all curves y. We can find $\mathcal{C}^1$ curves giving values of $J(y)$ arbitrarily close to 0, but cannot achieve $J(y) = 0$. On the other hand, the curve*

$$y(x) = \begin{cases} 0 & \text{if } -1 \leq x < 0, \\ x & \text{if } 0 \leq x \leq 1 \end{cases}$$

gives $J(y) = 0$. This curve is piecewise $\mathcal{C}^1$ with a corner point at $x = 0$. □

Suppose that a piecewise $\mathcal{C}^1$ curve y is a strong extremum for the Basic Calculus of Variations Problem, under the same assumptions as in Section 2.3. Clearly, y is then also a weak extremum, with respect to the generalized 1-norm

$$\|y\|_1 := \max_{a \leq x \leq b} |y(x)| + \max_{a \leq x \leq b} \max\{|y'(x^-)|, |y'(x^+)|\}. \tag{3.1}$$

As we stated in Section 2.3.3 (see in particular footnote 2 there), such an extremum must satisfy the integral form (2.23) of the Euler-Lagrange equation almost everywhere, i.e., at all noncorner points. Extending our previous terminology to the present setting, we will refer to piecewise $\mathcal{C}^1$ solutions of (2.23) as *extremals* (sometimes extremals with corner points are also called *broken extremals*). We now want to investigate what additional conditions must hold at corner points in order for y to be a strong extremum. We give a direct analysis[1] below; later we will mention an alternative way of deriving these conditions (see Exercise 3.3).

For simplicity, we assume that y has only one (unspecified) corner point $c \in [a, b]$. As a generalization of (2.10), we will let two separate perturbations η_1 and η_2 act on the two portions of y (before and after the corner point). To make this construction precise, denote these two portions by $y_1 : [a, c] \to \mathbb{R}$

---

[1]The reader who finds this derivation difficult to follow might wish to skip it at first reading. We also note that the insight gained from solving Exercise 2.6 should be helpful here.

and $y_2 : [c, b] \to \mathbb{R}$; their perturbed versions will then be $y_1 + \alpha \eta_1$ and $y_2 + \alpha \eta_2$. Clearly, we must have $\eta_1(a) = \eta_2(b) = 0$ to preserve the endpoints. Furthermore, since the location of the corner point is not fixed, we should allow the corner point of the perturbed curve to deviate from c. Let this new corner point be $c + \alpha \Delta x$ for some $\Delta x \in \mathbb{R}$, with the same α as before for convenience. Our family of perturbed curves (parameterized by α) is thus determined by the two curves η_1, η_2 and one real number Δx. We label these new curves as $y(\cdot, \alpha)$, with $y(\cdot, 0) = y$. There will be an additional condition on η_1 and η_2 to guarantee that $y(\cdot, \alpha)$ is continuous for each α; see (3.4) below. We take both η_1 and η_2 to be $\mathcal{C}^1$, to ensure that $y(\cdot, \alpha)$ is piecewise $\mathcal{C}^1$ with a single corner point at $c + \alpha \Delta x$. (Note that the difference $y(\cdot, \alpha) - y$ is piecewise $\mathcal{C}^1$ with two corner points, one at c and the other at $c + \alpha \Delta x$.) Figure 3.2 should help visualize this situation and the argument that follows.

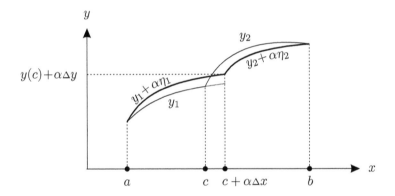

Figure 3.2: A perturbation of an extremal with a corner

The reader might have noticed a small problem which we need to fix before proceeding. The domain of y_1 is $[a, c]$, whereas we want the domain of $y_1 + \alpha \eta_1$ to be $[a, c + \alpha \Delta x]$. For $\alpha \Delta x > 0$, such a perturbed curve is ill defined. To deal with this issue, let us agree to extend y_1 beyond c via linear continuation: define y_1 for $x > c$ by $y_1(x) := y(c) + y'(c^-)(x - c)$. The linearity is actually not crucial, all we need is that the function y_1 be $\mathcal{C}^1$ at $x = c$, with

$$y_1(c) = y(c), \qquad y_1'(c) = y'(c^-). \tag{3.2}$$

If the perturbation η_1 is also defined on an interval extending to the right of c, then the earlier construction makes sense (at least for α close enough to 0). Of course we need to make a similar modification to y_2, extending it linearly to the left of c.

Let us write the functional to be minimized as a sum of two components:

$$
\begin{aligned}
J(y) &= \int_a^b L(x, y(x), y'(x))dx \\
&= \int_a^c L(x, y_1(x), y_1'(x))dx + \int_c^b L(x, y_2(x), y_2'(x))dx =: J_1(y_1) + J_2(y_2).
\end{aligned}
$$

After the perturbation, the first functional becomes

$$
J_1(y_1 + \alpha\eta_1) = \int_a^{c+\alpha\Delta x} L(x, y_1(x) + \alpha\eta_1(x), y_1'(x) + \alpha\eta_1'(x))dx
$$

(note that Δx should also be an argument on the left-hand side, but we omit it for simplicity). We can now compute the corresponding first variation:

$$
\begin{aligned}
\delta J_1|_{y_1}(\eta_1) &= \left.\frac{d}{d\alpha}\right|_{\alpha=0} J_1(y_1 + \alpha\eta_1) \\
&= \int_a^c \left(L_y(x, y_1(x), y_1'(x))\eta_1(x) + L_{y'}(x, y_1(x), y_1'(x))\eta_1'(x) \right)dx \\
&\quad + L(c, y_1(c), y_1'(c))\Delta x.
\end{aligned}
$$

Applying integration by parts and recalling (3.2) and the constraint $\eta_1(a) = 0$, we can bring the above expression to the form

$$
\begin{aligned}
\delta J_1|_{y_1}(\eta_1) &= \int_a^c \left(L_y(x, y_1(x), y_1'(x)) - \frac{d}{dx}L_{y'}(x, y_1(x), y_1'(x)) \right)\eta_1(x)dx \\
&\quad + L_{y'}(c, y(c), y'(c^-))\eta_1(c) + L(c, y(c), y'(c^-))\Delta x.
\end{aligned}
$$

Similarly, for the second functional we have

$$
J_2(y_2 + \alpha\eta_2) = \int_{c+\alpha\Delta x}^b L(x, y_2(x) + \alpha\eta_2(x), y_2'(x) + \alpha\eta_2'(x))dx
$$

and the first variation of J_2 at y_2 is

$$
\begin{aligned}
\delta J_2|_{y_2}(\eta_2) &= \int_c^b \left(L_y(x, y_2(x), y_2'(x)) - \frac{d}{dx}L_{y'}(x, y_2(x), y_2'(x)) \right)\eta_2(x)dx \\
&\quad - L_{y'}(c, y(c), y'(c^+))\eta_2(c) - L(c, y(c), y'(c^+))\Delta x.
\end{aligned}
$$

For α close to 0, the perturbed curve $y(\cdot, \alpha)$ is close to the original curve y in the sense of the 0-norm. Therefore, the function $\alpha \mapsto J(y(\cdot, \alpha))$ must attain a minimum at $\alpha = 0$, implying that

$$
\begin{aligned}
0 &= \left.\frac{d}{d\alpha}\right|_{\alpha=0} J(y(\cdot, \alpha)) = \left.\frac{d}{d\alpha}\right|_{\alpha=0} \left(J_1(y_1 + \alpha\eta_1) + J_2(y_2 + \alpha\eta_2) \right) \\
&= \delta J_1|_{y_1}(\eta_1) + \delta J_2|_{y_2}(\eta_2).
\end{aligned}
$$

Next, observe that each of the two portions y_i, $i = 1, 2$ of the optimal curve y must be an extremal of the corresponding functional J_i. Indeed, this becomes clear if we consider the special case when the perturbation η_i vanishes at c and $\Delta x = 0$. Therefore, the integrals in the preceding expressions for $\delta J_1|_{y_1}(\eta_1)$ and $\delta J_2|_{y_2}(\eta_2)$ should both vanish, and we are left with the condition

$$L_{y'}(c, y(c), y'(c^-))\eta_1(c) - L_{y'}(c, y(c), y'(c^+))\eta_2(c)$$
$$+ L(c, y(c), y'(c^-))\Delta x - L(c, y(c), y'(c^+))\Delta x = 0. \quad (3.3)$$

Now we need to take into account the fact that the two perturbations η_1 and η_2 are not independent: they have to be such that the perturbed curve remains continuous at $x = c + \alpha\Delta x$. This provides the additional relation

$$y_1(c + \alpha\Delta x) + \alpha\eta_1(c + \alpha\Delta x) = y_2(c + \alpha\Delta x) + \alpha\eta_2(c + \alpha\Delta x)$$
$$=: y(c) + \alpha\Delta y + o(\alpha). \quad (3.4)$$

The quantity Δy describes the first-order (in α) vertical displacement of the corner point, in much the same sense that Δx describes the first-order horizontal displacement; Δy and Δx are independent of each other. Equating the first-order terms with respect to α in (3.4) and using the second equality in (3.2) along with its counterpart $y_2'(c) = y'(c^+)$, we obtain

$$y'(c^-)\Delta x + \eta_1(c) = y'(c^+)\Delta x + \eta_2(c) = \Delta y. \quad (3.5)$$

Using (3.5), we can eliminate $\eta_1(c)$ and $\eta_2(c)$ from (3.3) and rewrite that formula in terms of Δy and Δx as follows:

$$\left(L_{y'}(c, y(c), y'(c^-)) - L_{y'}(c, y(c), y'(c^+))\right)\Delta y - \Big(\left(L_{y'}(c, y(c), y'(c^-))y'(c^-)\right.$$

$$\left.-L(c, y(c), y'(c^-))\right) - \left(L_{y'}(c, y(c), y'(c^+))y'(c^+) - L(c, y(c), y'(c^+))\right)\Big)\Delta x$$

$$= - \left.L_{y'}(x, y(x), y'(x))\right|_{c^-}^{c^+}\Delta y$$

$$+ \left.\left(L_{y'}(x, y(x), y'(x))y'(x) - L(x, y(x), y'(x))\right)\right|_{c^-}^{c^+}\Delta x = 0.$$

Since Δx and Δy are independent and arbitrary, we conclude that the terms multiplying them must be 0. This means that $L_{y'}$ and $y'L_{y'} - L$ are in fact continuous at $x = c$.

The above reasoning can be extended to multiple corner points, yielding the necessary conditions for optimality known as the **Weierstrass-Erdmann corner conditions**: *If a curve y is a strong extremum, then $L_{y'}$ and $y'L_{y'} - L$ must be continuous at each corner point of y.* More precisely, their discontinuities (due to the fact that y' does not exist at corner points) must be *removable*. The quantities $L_{y'}$ and $y'L_{y'} - L$ are of course familiar to us from Chapter 2; they are, respectively, the momentum and the Hamiltonian. Weierstrass presented these conditions in 1865 during his lectures on calculus of variations, but never formally published them. They were independently derived and published by Erdmann in 1877.

Exercise 3.1   *Let y be a weak extremum (with (3.1) serving as the 1-norm) but not a strong extremum. Carefully explain where the above proof breaks down. Show that, nevertheless, the first Weierstrass-Erdmann condition (the one that asserts continuity of $L_{y'}$) can still be established by suitably specializing the proof.*   □

For the case of a single corner point, the two Weierstrass-Erdmann corner conditions together with the two boundary conditions provide four relations, which is the correct number to uniquely specify two portions of the extremal (each satisfying the second-order Euler-Lagrange differential equation). In general, to uniquely specify an extremal consisting of m portions (i.e., having $m - 1$ corner points) we need $2m$ conditions, and these are provided by $2(m - 1)$ corner conditions plus the two boundary conditions.

Exercise 3.2   *Consider the problem of minimizing $J(y) = \int_{-1}^{1} (y'(x))^3 dx$ subject to the boundary conditions $y(-1) = y(1) = 0$. Characterize all piecewise $\mathcal{C}^1$ extremals of J. Do any of them satisfy the Weierstrass-Erdmann corner conditions? For each one that does, check if it is a minimum (weak or strong).*   □

3.1.2   Weierstrass excess function

To continue our search for additional conditions (besides being an extremal) which are necessary for a piecewise $\mathcal{C}^1$ curve y to be a strong minimum, we now introduce a new concept. For a given Lagrangian $L = L(x, y, z)$, the *Weierstrass excess function*, or *E-function*, is defined as

$$E(x, y, z, w) := L(x, y, w) - L(x, y, z) - (w - z) \cdot L_z(x, y, z). \qquad (3.6)$$

The above formula is written to suggest multiple degrees of freedom, but from now on we specialize to the single-degree-of-freedom case for simplicity. Note that $L(x, y, z) + (w - z)L_z(x, y, z)$ is the first-order Taylor approximation of $L(x, y, w)$, viewed as a function of w, around $w = z$. This gives the geometric interpretation of the E-function as the distance between the Lagrangian and its linear approximation around $w = z$; see Figure 3.3.

The **Weierstrass necessary condition for a strong minimum** states that *if $y(\cdot)$ is a strong minimum, then*

$$E(x, y(x), y'(x), w) \geq 0 \qquad (3.7)$$

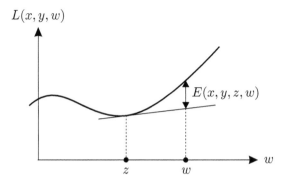

Figure 3.3: Weierstrass excess function

for all noncorner points $x \in [a, b]$ *and all* $w \in \mathbb{R}$. The geometric meaning of this condition is that for each x, the graph of the function $L(x, y(x), \cdot)$ lies above its tangent line at $y'(x)$, which can be interpreted as a local convexity property of this function.

The Weierstrass necessary condition can be proved as follows. Suppose that a curve y is a strong minimum. Let $\bar{x} \in [a, b]$ be a noncorner point of y, let $d \in (\bar{x}, b]$ be such that the interval $[\bar{x}, d]$ contains no corner points of y, and pick some $w \in \mathbb{R}$. We construct a family of perturbed curves $y(\cdot, \varepsilon)$, parameterized by $\varepsilon \in [0, d - \bar{x})$, which are continuous, coincide with y on the complement of $[\bar{x}, d]$, are linear with derivative w on $[\bar{x}, \bar{x} + \varepsilon]$, and differ from y by a linear function on $[\bar{x} + \varepsilon, d]$. The precise definition is

$$y(x, \varepsilon) := \begin{cases} y(x) & \text{if } a \leq x \leq \bar{x} \text{ or } d \leq x \leq b, \\ y(\bar{x}) + w(x - \bar{x}) & \text{if } \bar{x} \leq x \leq \bar{x} + \varepsilon, \\ y(x) + \dfrac{d - x}{d - (\bar{x} + \varepsilon)} \big(y(\bar{x}) + w\varepsilon - y(\bar{x} + \varepsilon)\big) & \text{if } \bar{x} + \varepsilon \leq x \leq d. \end{cases}$$
(3.8)

Such a perturbed curve $y(\cdot, \varepsilon)$ is shown in Figure 3.4.

It is clear that

$$y(\cdot, 0) = y \tag{3.9}$$

and that for ε close to 0 the perturbed curve $y(\cdot, \varepsilon)$ is close to the original curve y in the sense of the 0-norm. Therefore, the function $\varepsilon \mapsto J(y(\cdot, \varepsilon))$ must have a minimum at $\varepsilon = 0$. We will now show that the right-sided derivative of this function at $\varepsilon = 0$ exists and equals $E(\bar{x}, y(\bar{x}), y'(\bar{x}), w)$. Then, since this derivative must be nonnegative and $\bar{x}$ and w are arbitrary, the proof will be complete.

Noting that the behavior of $y(\cdot, \varepsilon)$ outside the interval $[\bar{x}, d]$ does not

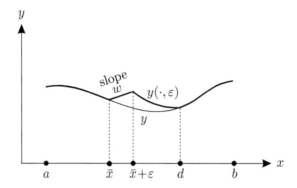

Figure 3.4: The graphs of y and $y(\cdot, \varepsilon)$

depend on ε, we have

$$
\frac{d}{d\varepsilon}J(y(\cdot, \varepsilon)) = \frac{d}{d\varepsilon}\left(\int_{\bar{x}}^{\bar{x}+\varepsilon} L(x, y(x, \varepsilon), y_x(x, \varepsilon))dx \right.
$$
$$
\left. + \int_{\bar{x}+\varepsilon}^{d} L(x, y(x, \varepsilon), y_x(x, \varepsilon))dx \right). \tag{3.10}
$$

By (3.8), the first integral in (3.10) is $\int_{\bar{x}}^{\bar{x}+\varepsilon} L(x, y(\bar{x}) + w(x - \bar{x}), w)dx$ and its derivative is simply

$$
L(\bar{x} + \varepsilon, y(\bar{x}) + w\varepsilon, w). \tag{3.11}
$$

The differentiation of the second integral in (3.10) gives

$$
- L(\bar{x} + \varepsilon, y(\bar{x} + \varepsilon, \varepsilon), y_x(\bar{x} + \varepsilon, \varepsilon)) + \int_{\bar{x}+\varepsilon}^{d} L_y(x, y(x, \varepsilon), y_x(x, \varepsilon))y_\varepsilon(x, \varepsilon)dx
$$
$$
+ \int_{\bar{x}+\varepsilon}^{d} L_{y'}(x, y(x, \varepsilon), y_x(x, \varepsilon))y_{x\varepsilon}(x, \varepsilon)dx \tag{3.12}
$$

(it is straightforward to check that the partial derivatives $y_x(x, \varepsilon)$, $y_\varepsilon(x, \varepsilon)$, $y_{x\varepsilon}(x, \varepsilon)$ exist inside the relevant intervals). Since $y_{x\varepsilon}(x, \varepsilon) = y_{\varepsilon x}(x, \varepsilon) = \frac{d}{dx}y_\varepsilon(x, \varepsilon)$, we can use integration by parts to bring the last integral in (3.12) to the form

$$
L_{y'}(x, y(x, \varepsilon), y_x(x, \varepsilon))y_\varepsilon(x, \varepsilon)\big|_{\bar{x}+\varepsilon}^{d}
$$
$$
- \int_{\bar{x}+\varepsilon}^{d} \frac{d}{dx}\Big(L_{y'}(x, y(x, \varepsilon), y_x(x, \varepsilon)) \Big) y_\varepsilon(x, \varepsilon)dx. \tag{3.13}
$$

In more detail, the first term in (3.13) is

$$
L_{y'}(d, y(d, \varepsilon), y_x(d, \varepsilon))y_\varepsilon(d, \varepsilon) - L_{y'}(\bar{x}+\varepsilon, y(\bar{x}+\varepsilon, \varepsilon), y_x(\bar{x}+\varepsilon, \varepsilon))y_\varepsilon(\bar{x}+\varepsilon, \varepsilon). \tag{3.14}
$$

We have from (3.8) that $y(d, \varepsilon) = y(d)$ for all ε, hence $y_\varepsilon(d, \varepsilon) = 0$ and so the first term in (3.14) is 0. Another consequence of (3.8) is the relation $y(\bar{x} + \varepsilon, \varepsilon) = y(\bar{x}) + w\varepsilon$. Differentiating it with respect to ε, we obtain $y_x(\bar{x} + \varepsilon, \varepsilon) + y_\varepsilon(\bar{x} + \varepsilon, \varepsilon) = w$, or $y_\varepsilon(\bar{x} + \varepsilon, \varepsilon) = w - y_x(\bar{x} + \varepsilon, \varepsilon)$. When we substitute this expression for $y_\varepsilon(\bar{x} + \varepsilon, \varepsilon)$ into the second term in (3.14), that term becomes

$$-L_{y'}(\bar{x} + \varepsilon, y(\bar{x} + \varepsilon, \varepsilon), y_x(\bar{x} + \varepsilon, \varepsilon))(w - y_x(\bar{x} + \varepsilon, \varepsilon)). \tag{3.15}$$

We see that $\frac{d}{d\varepsilon} J(y(\cdot, \varepsilon))$ is given by the sum of (3.11), the first two terms in (3.12), the integral (with the minus sign) in (3.13), and (3.15). Setting $\varepsilon = 0$ and rearranging terms, we arrive at

$$\frac{d}{d\varepsilon}\Big|_{\varepsilon=0^+} J(y(\cdot, \varepsilon)) = L(\bar{x}, y(\bar{x}), w) - L(\bar{x}, y(\bar{x}, 0), y_x(\bar{x}, 0))$$
$$- L_{y'}(\bar{x}, y(\bar{x}, 0), y_x(\bar{x}, 0))(w - y_x(\bar{x}, 0))$$
$$+ \int_{\bar{x}}^{d} \left(L_y(x, y(x,0), y_x(x, 0)) - \frac{d}{dx} L_{y'}(x, y(x, 0), y_x(x, 0)) \right) y_\varepsilon(x, 0) dx.$$

Now, recall (3.9) which also implies that $y_x(\cdot, 0) = y'$. Thus the integral in the previous formula equals

$$\int_{\bar{x}}^{d} \left(L_y(x, y(x), y'(x)) - \frac{d}{dx} L_{y'}(x, y(x), y'(x)) \right) y_\varepsilon(x, 0) dx = 0$$

because y as a strong (hence also weak) minimum must satisfy the Euler-Lagrange equation. We are left with

$$\frac{d}{d\varepsilon}\Big|_{\varepsilon=0^+} J(y(\cdot, \varepsilon)) = L(\bar{x}, y(\bar{x}), w) - L(\bar{x}, y(\bar{x}), y'(\bar{x}))$$
$$- L_{y'}(\bar{x}, y(\bar{x}), y'(\bar{x}))(w - y'(\bar{x})) = E(\bar{x}, y(\bar{x}), y'(\bar{x}), w)$$

as desired, and the Weierstrass necessary condition is established. A necessary condition for a strong *maximum* is analogous but with the reversed inequality sign; this can be verified by passing from L to $-L$ or by modifying the proof in the obvious way. The Weierstrass necessary condition can also be extended to corner points, either by refining the proof or via a limiting argument (cf. Exercise 3.3 below).

Weierstrass introduced the above necessary condition during his 1879 lectures on calculus of variations. His original proof relied on an additional assumption (normality) which was subsequently removed by McShane in 1939. Let us now take a few moments to reflect on the perturbation used in the proof we just gave. First, it is important to observe that the perturbed curve $y(\cdot, \varepsilon)$ is close to the original curve y in the sense of the 0-norm, but not necessarily in the sense of the 1-norm. Indeed, it is clear that

$\|y(\cdot, \varepsilon) - y\|_0 \to 0$ as $\varepsilon \to 0$; on the other hand, the derivative of $y(\cdot, \varepsilon)$ for x immediately to the right of $\bar{x}$ equals w, hence $\|y(\cdot, \varepsilon) - y\|_1 \geq |w - y'(\bar{x})|$ for all $\varepsilon > 0$, no matter how small. For this reason, the necessary condition applies only to strong minima, unless we restrict w to be sufficiently close to $y'(\bar{x})$. Note also that the first variation was not used in the proof. Thus we have already departed significantly from the variational approach which we followed in Chapter 2. Derived using a richer class of perturbations, the Weierstrass necessary condition turns out to be powerful enough to yield as its corollaries the Weierstrass-Erdmann corner conditions from Section 3.1.1 as well as Legendre's condition from Section 2.6.1.

Exercise 3.3   *Use the Weierstrass necessary condition to prove that a piecewise $\mathcal{C}^1$ strong minimum must satisfy the Weierstrass-Erdmann corner conditions and Legendre's condition.*   □

When solving this exercise, the reader should keep the following points in mind. First, we know from Exercise 3.1 that the first Weierstrass-Erdmann corner condition is necessary for weak extrema as well. This condition should thus follow directly from the fact that y is an extremal—i.e., satisfies the integral form (2.23) of the Euler-Lagrange equation—without the need to apply the Weierstrass necessary condition. The second Weierstrass-Erdmann corner condition, on the other hand, is necessary only for strong extrema, and deducing it requires the full power of the Weierstrass necessary condition (including a further analysis of what the latter implies for corner points). As for Legendre's condition, it can be derived from the local version of the Weierstrass necessary condition with w restricted to be close to $y'(x)$ for a given x, thus confirming that Legendre's condition is also necessary for weak extrema. Finally, when x is a corner point, Legendre's condition should read $L_{y'y'}(x, y(x), y'(x^{\pm})) \geq 0$.

The perturbation used in the above proof of the Weierstrass necessary condition is already quite close to the ones we will use later in the proof of the maximum principle. The main difference is that in the proof of the maximum principle, we will not insist on bringing the perturbed curve back to the original curve after the perturbation stops acting. Instead, we will analyze how the effect of a perturbation applied on a small interval propagates up to the terminal point.

There is a very insightful reformulation of the Weierstrass necessary condition which reveals its direct connection to the Hamiltonian maximization property discussed at the end of Section 2.6.1 (and thus to the maximum principle which we are steadily approaching). Let us write our Hamiltonian (2.63) as

$$H(x, y, z, p) = zp - L(x, y, z).$$

Then a simple manipulation of (3.6) allows us to bring the E-function and the condition (3.7) to the form

$$
\begin{aligned}
E(x, y(x), y'(x), w) &= y'(x) L_z(x, y(x), y'(x)) - L(x, y(x), y'(x)) \\
&\quad - \left(w L_z(x, y(x), y'(x)) - L(x, y(x), w) \right) \\
&= H(x, y(x), y'(x), p(x)) - H(x, y(x), w, p(x)) \geq 0
\end{aligned}
$$

where we used the formula

$$
p(x) = L_z(x, y(x), y'(x))
$$

consistent with our earlier definition of the momentum (see Section 2.4.1). Therefore, the Weierstrass necessary condition simply says that if $y(\cdot)$ is an optimal trajectory and $p(\cdot)$ is the corresponding momentum, then for every x the function $H(x, y(x), \cdot, p(x))$, which is the same as the function H^* defined in (2.32), has a maximum at $y'(x)$. This interpretation escaped Weierstrass, not just because it requires bringing in the Hamiltonian but because it demands treating z and p as independent arguments of the Hamiltonian (we already discussed this point at the end of Section 2.4.2).

Combining the Weierstrass necessary condition with the sufficient condition for a weak minimum from Section 2.6.2, one can obtain a sufficient condition for a strong minimum. The precise formulation of this condition requires a new concept (that of a field) and we will not develop it. While sufficient conditions for optimality are theoretically appealing, they tend to be less practical to apply compared to necessary conditions; we already saw this in Section 2.6.2 and will see it again in Chapter 5.

3.2 CALCULUS OF VARIATIONS VERSUS OPTIMAL CONTROL

Problems in calculus of variations that we have treated so far are concerned with minimizing a cost functional of the form $J(y) = \int_a^b L(x, y(x), y'(x)) dx$ over a given family of curves $y(\cdot)$—such as, e.g., all $\mathcal{C}^1$ curves with fixed endpoints. Optimal control theory studies similar problems but from a more *dynamic* viewpoint, which can be explained as follows. Rather than regarding the curves as given a priori, let us imagine a particle moving in the (x, y)-space and "drawing a trace" of its motion. The choice of the slope $y'(x)$ at each point on the curve can be thought of as an infinitesimal decision, or control. The resulting curve is thus a trajectory of a simple control system, which we can write as $y' = u$. In order for this curve to minimize the overall integral cost, optimal control decisions must be taken everywhere along the curve; this is simply a restatement of a principle that we have already discussed several times in Chapter 2 (see, in particular, pages 38 and 50).

In realistic scenarios, not all velocities may be feasible everywhere. In calculus of variations, constraints on available velocities may be modeled as equalities of the form $M(x, y(x), y'(x)) = 0$. We already know from Section 2.5.2 that if we solve such a constraint for y' and parameterize the solution in terms of free variables u, we arrive at a control system $y' = f(x, y, u)$. This dynamic description is consistent with the idea of moving along the curve (and incurring a cost along the way).

The set in which the controls u take values might also be constrained by some practical considerations, such as inherent bounds on physical quantities (velocities, forces, and so on). In the optimal control formulation, such constraints are incorporated very naturally by working with an appropriate control set. In calculus of variations, on the other hand, they would make the description of the space of admissible curves quite cumbersome.

Finally, once we adopt the dynamic viewpoint of a moving particle, it is natural to consider another transformation which we already encountered in Section 2.4.3. Namely, it makes sense to parameterize the curves by *time* rather than by the spatial variable x. Besides being more intuitive, this new formulation is also more descriptive because it allows us to distinguish between two geometrically identical curves traversed with different speeds. In addition, the curves no longer need to be graphs of single-valued functions of x.

Example 3.2   *The brachistochrone problem provides a great example in which the two formulations—the earlier one of calculus of variations and the new one of optimal control—are both meaningful and can be clearly compared. Our old formulation, given in Section 2.1.4, was in terms of minimizing the cost functional (2.6). The Lagrangian inside that integral is the ratio of the arclength to the speed, where we used the fact that the speed satisfies the equation (2.5) as a consequence of the conservation of energy law (2.4).*

Let us now adopt a different approach and parameterize the curves by time. In other words, represent each point on a given curve as $(x(t), y(t))$, where the x-axis and the y-axis are as shown in Figure 2.4 and t is the time at which the particle arrives at this point; we assume that at time t_0 the particle is resting at the initial point $(a, 0)$. The speed constraint (2.5) then becomes $\dot{x}^2 + \dot{y}^2 = y$. Defining $u_1 := \dot{x}/\sqrt{y}$ and $u_2 := \dot{y}/\sqrt{y}$, we can describe the motion by the control system

$$\dot{x} = u_1 \sqrt{y},$$
$$\dot{y} = u_2 \sqrt{y}$$

subject to the constraint $u_1^2 + u_2^2 = 1$ (i.e., the control set is the unit circle).

Now, the quantity to be minimized is the time it takes for the particle to arrive at a specified destination (b, y_1). If the curve is parameterized by

$t \in [t_0, t_1]$, *then this is simply*

$$J = t_1 - t_0 = \int_{t_0}^{t_1} 1 dt. \tag{3.16}$$

Thus the Lagrangian is $L \equiv 1$. Note that we avoided writing $J(y)$ since we are no longer working with a curve $y = y(x)$; in fact, it makes more sense to write $J(u_1, u_2)$ because the curve is determined by the choice of controls as functions of time.

The new problem formulation is equivalent to the original one, except that it is more general in one aspect: as we already mentioned, it avoids the (sometimes tacit) assumption made in calculus of variations that admissible curves are graphs of functions of x. Observe that in terms of complexity of the problem description, the burden has shifted from the cost functional to the right-hand side of the control system. □

From this point onward, we will start using t as the independent variable. We will write $x = (x_1, \ldots, x_n)^T$ for the (dependent) state variables, $\dot{x} = (\dot{x}_1, \ldots, \dot{x}_n)^T$ for their time derivatives, and $u = (u_1, \ldots, u_m)^T$ for the controls. The controls will take values in some control set, such as the unit circle in the above example. Of course, the simplicity of the Lagrangian in (3.16) is due to the fact that the cost being minimized is the time (this is a *time-optimal* control problem); in general, both the control system and the cost functional may be quite complex.

3.3 OPTIMAL CONTROL PROBLEM FORMULATION AND ASSUMPTIONS

Having completed our study of calculus of variations, we have now come full circle and are ready to tackle the optimal control problem posed at the very beginning of the book, in Section 1.1. The purpose of this section is to give a more detailed formulation of that problem, by discussing and comparing its several specific variants and spelling out necessary technical assumptions.

3.3.1 Control system

In the notation announced at the end of the previous section, control systems that we want to study take the form

$$\dot{x} = f(t, x, u), \qquad x(t_0) = x_0 \tag{3.17}$$

which is an exact copy of (1.1). Here $x \in \mathbb{R}^n$ is the state, $u \in U \subset \mathbb{R}^m$ is the control, $t \in \mathbb{R}$ is the time, t_0 is the initial time, and x_0 is the initial state. Both x and u are functions of time: $x = x(t)$, $u = u(t)$. The control set U is usually a closed subset of $\mathbb{R}^m$ and can be the entire $\mathbb{R}^m$; in principle it can also vary with time, but here we take it to be fixed.

We want to know that for every choice of the initial data (t_0, x_0) and every admissible control $u(\cdot)$, the system (3.17) has a unique solution $x(\cdot)$ on some time interval $[t_0, t_1]$. If this property holds, we will say that the system is *well posed*. To guarantee local existence and uniqueness of solutions for (3.17), we need to impose some regularity conditions on the right-hand side f and on the admissible controls u. For the sake of simplicity, we will usually be willing to make slightly stronger assumptions than necessary; when we do this, we will briefly indicate how our assumptions can be relaxed.[2]

Let us begin by considering the case of no controls:

$$\dot{x} = f(t, x). \tag{3.18}$$

First, to assure sufficient regularity of f with respect to t, we take $f(\cdot, x)$ to be piecewise continuous for each fixed x. Here, by a *piecewise continuous* function we mean a function having at most a finite number of discontinuities on every bounded interval, and possessing finite limits from the right and from the left at each of these discontinuities. For convenience, we assume that the value of such a function at each discontinuity is equal to one of these one-sided limits (i.e., the function is either left-continuous or right-continuous at each point). The assumption of a finite number of discontinuities on each bounded interval is actually not crucial; we can allow discontinuities to have accumulation points, as long as the function remains locally bounded (or at least locally integrable).

Second, we need to specify how regular f should be with respect to x. A standard assumption in this regard is that f is locally Lipschitz in x, uniformly over t. Namely, for every (t_0, x_0) there should exist a constant L such that we have

$$|f(t, x_1) - f(t, x_2)| \leq L|x_1 - x_2|$$

for all (t, x_1) and (t, x_2) in some neighborhood of (t_0, x_0) in $\mathbb{R} \times \mathbb{R}^n$. We can in fact be more generous and assume the following: $f(t, \cdot)$ is $\mathcal{C}^1$ for each fixed t, and $f_x(\cdot, x)$ is piecewise continuous for each fixed x. It is easy to verify using the Mean Value Theorem that such a function f satisfies the previous Lipschitz condition. Note that here and below, we extend our earlier notation $f_x := \partial f/\partial x$ to the vector case, so that f_x stands for the Jacobian matrix of partial derivatives of f with respect to x.

If f satisfies the above regularity assumptions, then on some interval $[t_0, t_1]$ there exists a unique solution $x(\cdot)$ of the system (3.18). Since we did

---

[2]The reader not interested in studying these technical conditions at the present time may skip forward to Section 3.3.2 with no significant loss of continuity.

not assume that f is continuous with respect to t, some care is needed in interpreting what we mean by a solution of (3.18). In the present situation, it is reasonable to call a function $x(\cdot)$ a solution of (3.18) if it is continuous everywhere, $\mathcal{C}^1$ almost everywhere, and satisfies the corresponding integral equation

$$x(t) = x_0 + \int_{t_0}^t f(s, x(s))ds.$$

A function $x(\cdot)$ that can be represented as an integral of another function $g(\cdot)$, and thus automatically satisfies $\dot{x} = g$ almost everywhere, is called *absolutely continuous*. This class of functions generalizes the piecewise $\mathcal{C}^1$ functions that we considered earlier. Basically, the extra generality here is that the derivative can be discontinuous on a set of points that has measure zero (e.g., a countable set) rather than at a finite number of points on a bounded interval, and can approach infinity near these points. If we insist that the derivative be locally bounded, we arrive at the slightly smaller class of locally Lipschitz functions.

We are now ready to go back to the control system (3.17). To guarantee local existence and uniqueness of its solutions, we can impose assumptions on f and u that would let us invoke the previous existence and uniqueness result for the right-hand side

$$\bar{f}(t, x) := f(t, x, u(t)). \tag{3.19}$$

Here is one such set of assumptions which, although not the weakest possible, is adequate for our purposes: f is continuous in t and u and $\mathcal{C}^1$ in x; f_x is continuous in t and u; and $u(\cdot)$ is piecewise continuous as a function of t. Another, weaker set of hypotheses is obtained by replacing the assumptions of existence of f_x and its continuity with respect to all variables with the following Lipschitz property: for every bounded subset D of $\mathbb{R} \times \mathbb{R}^n \times U$, there exists an L such that we have

$$|f(t, x_1, u) - f(t, x_2, u)| \le L|x_1 - x_2|$$

for all $(t, x_1, u), (t, x_2, u) \in D$. Note that in either case, differentiability of f with respect to u is not assumed.

Exercise 3.4 *Verify that each of the two sets of hypotheses just described guarantees local existence and uniqueness of solutions of (3.17), by applying the result stated earlier to the system with no controls defined by (3.19). Explain which hypotheses can be further relaxed.* $\square$

When the right-hand side does not explicitly depend on time, i.e., when we have $f = f(x, u)$, a convenient way to guarantee that the above conditions hold is to assume that f is locally Lipschitz (as a function from $\mathbb{R}^n \times U$ to $\mathbb{R}^n$).

In general, f depends on t in two ways: directly through the t-argument, and indirectly through u. Regarding the first dependence, we are willing to be generous by assuming that f is at least continuous in t. In fact, we can always eliminate the direct dependence of f on t by introducing the extra state variable $x_{n+1} := t$, with the dynamics $\dot{x}_{n+1} = 1$. Note that in order for the new system obtained in this way to satisfy our conditions for existence and uniqueness of solutions, continuity of f in t is not enough and we need f to be $\mathcal{C}^1$ in t (or satisfy an appropriate Lipschitz condition).

On the other hand, as far as regularity of u with respect to t is concerned, it would be too restrictive to assume anything stronger than piecewise continuity. In fact, occasionally we may even want to relax this assumption and allow u to be a locally bounded function with countably many discontinuities. More precisely, the class of admissible controls can be defined to consist of functions u that are measurable[3] and locally bounded. In view of the remarks made earlier about the system (3.18), local existence and uniqueness of solutions is still guaranteed for this larger class of controls. We will rarely need this level of generality, and piecewise continuous controls will be adequate for most of our purposes (with the exception of some of the material to be discussed in Sections 4.4 and 4.5—specifically, Fuller's problem and Filippov's theorem). Similarly to the case of the system (3.18), by a solution of (3.17) we mean an absolutely continuous function $x(\cdot)$ satisfying

$$x(t) = x_0 + \int_{t_0}^{t} f(s, x(s), u(s))ds.$$

In what follows, we will always assume that the property of local existence and uniqueness of solutions holds for a given control system. This of course does not guarantee that solutions exist globally in time. Typically, we will consider a candidate optimal trajectory defined on some time interval $[t_0, t_1]$, and then existence over the same time interval will be automatically ensured for nearby trajectories.

3.3.2   Cost functional

We will consider cost functionals of the form

$$J(u) := \int_{t_0}^{t_f} L(t, x(t), u(t))dt + K(t_f, x_f) \qquad (3.20)$$

which is an exact copy of (1.2). Here t_f and $x_f := x(t_f)$ are the *final* (or *terminal*) time and state, $L : \mathbb{R} \times \mathbb{R}^n \times U \to \mathbb{R}$ is the *running cost* (or Lagrangian), and $K : \mathbb{R} \times \mathbb{R}^n \to \mathbb{R}$ is the *terminal cost*. We will explain how

---

[3]The class of *measurable* functions is obtained from that of piecewise continuous functions by taking the closure with respect to almost-everywhere convergence.

the final time t_f is defined in Section **3.3.3** below. Since the cost depends on the initial data and the final time as well as on the control, it would be more accurate to write $J(t_0, x_0, t_f, u)$, but we write $J(u)$ for simplicity and to reflect the fact that the cost is being minimized over the space of control functions. Note that even if L does not depend on u, the cost J depends on the control $u(\cdot)$ through $x(\cdot)$ which is the trajectory that this control generates. The reader might have remarked that our present choice of arguments for L deviates from the one we made in calculus of variations; it seems that $L(t, x, \dot{x})$ would have been more consistent. However, we can always pass from $(t, x, \dot{x})$ to (t, x, u) by substituting $f(t, x, u)$ for $\dot{x}$, while it may not be possible to go in the opposite direction. Thus it is more general, as well as more natural, to let L depend explicitly on u. In contrast with Section **3.3.1**, where the regularity conditions on f and u were dictated by the goal of having a well-posed control system, there are no such a priori requirements on the functions L and K. All derivatives that will appear in our subsequent derivations will be assumed to exist, and depending on the analysis method we will eventually see what differentiability assumptions on L and K are needed.

Optimal control problems in which the cost is given by (3.20) are known as problems in the *Bolza form*, or collectively as the *Bolza problem*. There are two important special cases of the Bolza problem. The first one is the *Lagrange problem*, in which there is no terminal cost: $K \equiv 0$. This problem—and its name—of course come from calculus of variations. The second special case is the *Mayer problem*, in which there is no running cost: $L \equiv 0$. We can pass back and forth between these different forms by means of simple transformations. Indeed, given a problem with a terminal cost K, we can write

$$K(t_f, x_f) = K(t_0, x_0) + \int_{t_0}^{t_f} \frac{d}{dt} K(t, x(t)) dt$$

$$= K(t_0, x_0) + \int_{t_0}^{t_f} \big(K_t(t, x(t)) + K_x(t, x(t)) \cdot f(t, x(t), u(t))\big) dt.$$

Since $K(t_0, x_0)$ is a constant independent of u, we arrive at an equivalent problem in the Lagrange form with $K_t(t, x) + K_x(t, x) \cdot f(t, x, u)$ added to the original running cost. On the other hand, given a problem with a running cost L satisfying the same regularity conditions as f, we can introduce an extra state variable x^0 via

$$\dot{x}^0 = L(t, x, u), \qquad x^0(t_0) = 0$$

(we use a superscript instead of a subscript to avoid confusion with the initial state). This yields

$$\int_{t_0}^{t_f} L(t, x(t), u(t)) dt = x^0(t_f)$$

thus converting the problem to the Mayer form. (The value of $x^0(t_0)$ can actually be arbitrary, since it only changes the cost by an additive constant.) Note that the similar trick of introducing the additional state variable $x_{n+1} := t$, which we already mentioned in Section 3.3.1, eliminates the dependence of L and/or K on time; for the Bolza problem this gives $J(u) = \int_{t_0}^{t_f} L(x(t), u(t))dt + K(x_f)$ with $x \in \mathbb{R}^{n+1}$.

3.3.3   Target set

As we noted before, the cost functional (3.20) depends on the choice of t_0, x_0, and t_f. We take the initial time t_0 and the initial state x_0 to be fixed as part of the control system (3.17). We now need to explain how to define the final time t_f (which in turn determines the corresponding final state x_f). Depending on the control objective, the final time and final state can be free or fixed, or can belong to some set. All these possibilities are captured by introducing a *target set* $S \subset [t_0, \infty) \times \mathbb{R}^n$ and letting t_f be the smallest time such that $(t_f, x_f) \in S$. It is clear that t_f defined in this way in general depends on the choice of the control u. We will take S to be a closed set; hence, if $(t, x(t))$ ever enters S, the time t_f is well defined. If a trajectory is such that $(t, x(t))$ does not belong to S for any t, then we consider its cost as being infinite (or undefined). Note that here we do not allow the option that $t_f = \infty$ may give a valid finite cost, although we will study such "infinite-horizon" problems later (in Chapters 5 and 6). Below are some examples of target sets that we will encounter in the sequel.

The target set $S = [t_0, \infty) \times \{x_1\}$, where x_1 is a fixed point in $\mathbb{R}^n$, gives a *free-time, fixed-endpoint problem*. A generalization of this is to consider a target set of the form $S = [t_0, \infty) \times S_1$, where S_1 is a surface (manifold) in $\mathbb{R}^n$. Another natural target set is $S = \{t_1\} \times \mathbb{R}^n$, where t_1 is a fixed time in $[t_0, \infty)$; this gives a *fixed-time, free-endpoint problem*. It is useful to observe that if we start with a fixed-time, free-endpoint problem and consider again the auxiliary state $x_{n+1} := t$, we recover the previous case with $S_1 \subset \mathbb{R}^{n+1}$ given by $\{x \in \mathbb{R}^{n+1} : x_{n+1} = t_1\}$. A target set $S = T \times S_1$, where T is some subset of $[t_0, \infty)$ and S_1 is some surface in $\mathbb{R}^n$, includes as special cases all the target sets mentioned above. It also includes target sets of the form $S = \{t_1\} \times \{x_1\}$, which corresponds to the most restrictive case of a *fixed-time, fixed-endpoint problem*. As the opposite extreme, we can have $S = [t_0, \infty) \times \mathbb{R}^n$, i.e., a *free-time, free-endpoint problem*. (The reader may wonder about the sensibility of this latter problem formulation: when will the motion stop, or why would it even begin in the first place? To answer these questions, we have to keep in mind that the control objective is to minimize the cost (3.20). In the case of the Mayer problem, for example, (t_f, x_f) will be a point where the terminal cost is minimized, and if this minimum is unique then we do not need to specify a target set a priori. In the presence of a running cost L taking both positive and negative values, it is clear that remaining at rest at the initial state may not be optimal—and we also know that we can always bring such a problem to the Mayer form. When one says "cost" one may often think of it implicitly as a positive quantity, but remember that this need not be the case; we may be making a "profit" instead.) Many other target sets can be envisioned. For example, $S = \{(t, g(t)) : t \in [t_0, \infty)\}$ for some continuous function $g : \mathbb{R} \to \mathbb{R}^n$ corresponds to hitting a moving target. A point target can be generalized to a set by making g set-valued. The familiar trick of incorporating time as an extra state variable allows us to reduce such target sets to the ones we already discussed, and so we will not specifically consider them.

We now have a refined formulation of the optimal control problem: Given a control system (3.17) satisfying the assumptions of Section 3.3.1, a cost functional given by (3.20), and a target set $S \subset [t_0, \infty) \times \mathbb{R}^n$, find a control $u(\cdot)$ that minimizes the cost. Unlike in calculus of variations, we will usually interpret optimality in the global sense. (The necessary conditions for optimality furnished by the maximum principle apply to locally optimal controls as well, provided that we work with an appropriate norm; see Section 4.3 for details.)

3.4 VARIATIONAL APPROACH TO THE FIXED-TIME, FREE-ENDPOINT PROBLEM

We now want to see how far a variational approach—i.e., an approach based, as in Chapter 2, on analyzing the first (and second) variation of the cost functional—can take us in studying the optimal control problem formulated in Section 3.3. We will find that with only a modest amount of extra work, this familiar approach allows us to arrive at the correct statement of the maximum principle for a specific problem setting. We will also discover that, in general, this approach has serious shortcomings which will force us to turn our attention to a different, richer class of perturbations.

3.4.1 Preliminaries

Consider the optimal control problem from Section 3.3 with the following additional specifications: the target set is $S = \{t_1\} \times \mathbb{R}^n$, where t_1 is a fixed time (so this is a fixed-time, free-endpoint problem); $U = \mathbb{R}^m$ (the control is unconstrained); and the terminal cost is $K = K(x_f)$, with no direct dependence on the final time (just for simplicity). We can rewrite the cost in terms of the fixed final time t_1 as

$$J(u) = \int_{t_0}^{t_1} L(t, x(t), u(t))dt + K(x(t_1)). \tag{3.21}$$

Our goal is to derive necessary conditions for optimality. Let $u^*(\cdot)$ be an optimal control, by which we presently mean that it provides a global minimum: $J(u^*) \leq J(u)$ for all piecewise continuous controls u. Let $x^*(\cdot)$ be the corresponding optimal trajectory. We would like to consider nearby trajectories of the familiar form

$$x = x^* + \alpha\eta \tag{3.22}$$

but we must make sure that these perturbed trajectories are still solutions of the system (3.17), for suitably chosen controls. Unfortunately, the class of perturbations η that are admissible in this sense is difficult to characterize if we start with (3.22). Note also that the cost J, whose first variation we will be computing, is a function of u and not of x. Thus, in the optimal control context it is more natural to *directly perturb the control* instead, and then define perturbed state trajectories in terms of perturbed controls. To this end, we consider controls of the form

$$u = u^* + \alpha\xi \tag{3.23}$$

where ξ is a piecewise continuous function from $[t_0, t_1]$ to $\mathbb{R}^m$ and α is a real parameter as usual. We now want to find (if possible) a function $\eta : [t_0, t_1] \to \mathbb{R}^n$ for which the solutions of (3.17) corresponding to the controls (3.23), for a fixed ξ, are given by (3.22). Actually, we do not have any reason to believe that the perturbed trajectory depends linearly on α. Thus we should replace (3.22) by the more general (and more realistic) expression

$$x = x^* + \alpha\eta + o(\alpha). \tag{3.24}$$

It is obvious that $\eta(t_0) = 0$ since the initial condition does not change. Next, we derive a differential equation for η. Let us use the more detailed notation $x(t, \alpha)$ for the solution of (3.17) at time t corresponding to the control (3.23). The function $x(\cdot, \alpha)$ coincides with the right-hand side of (3.24) if and only if

$$x_\alpha(t, 0) = \eta(t) \tag{3.25}$$

for all t. (We are assuming here that the partial derivative x_α exists, but its existence can be shown rigorously; cf. Section 4.2.4.) Differentiating the quantity (3.25) with respect to time and interchanging the order of partial derivatives, we have

$$\dot{\eta}(t) = \frac{d}{dt}x_\alpha(t, 0) = x_{\alpha t}(t, 0) = x_{t\alpha}(t, 0) = \frac{d}{d\alpha}\bigg|_{\alpha=0} \dot{x}(t, \alpha)$$

$$= \frac{d}{d\alpha}\bigg|_{\alpha=0} f(t, x(t, \alpha), u^*(t) + \alpha\xi(t))$$

$$= f_x(t, x(t, 0), u^*(t))x_\alpha(t, 0) + f_u(t, x(t, 0), u^*(t))\xi(t)$$

$$= f_x(t, x^*(t), u^*(t))\eta(t) + f_u(t, x^*(t), u^*(t))\xi(t)$$

which we write more compactly (remembering also the initial condition $\eta(t_0) = 0$) as

$$\dot{\eta} = f_x(t, x^*, u^*)\eta + f_u(t, x^*, u^*)\xi =: f_x|_* \eta + f_u|_* \xi, \qquad \eta(t_0) = 0. \quad (3.26)$$

Here and below, we use the shorthand notation $|_*$ to indicate that a function is being evaluated along the optimal trajectory. The linear time-varying system (3.26) is nothing but the linearization of the original system (3.17) around the optimal trajectory. To emphasize the linearity of the system (3.26) we can introduce the notation $A_*(t) := f_x|_* (t)$ and $B_*(t) := f_u|_* (t)$ for the matrices appearing in it, bringing it to the form

$$\dot{\eta} = A_*(t)\eta + B_*(t)\xi, \qquad \eta(t_0) = 0. \quad (3.27)$$

The optimal control u^* minimizes the cost given by (3.21), and the control system (3.17) can be viewed as imposing the pointwise-in-time (non-integral) constraint $\dot{x}(t) - f(t, x(t), u(t)) = 0$. Motivated by Lagrange's idea for treating such constraints in calculus of variations, expressed by the augmented cost (2.53) on page 56, let us rewrite our cost as

$$J(u) = \int_{t_0}^{t_1} \big(L(t, x(t), u(t)) + p(t) \cdot (\dot{x}(t) - f(t, x(t), u(t)))\big) dt + K(x(t_1))$$

for some $\mathcal{C}^1$ function $p : [t_0, t_1] \to \mathbb{R}^n$ to be selected later. Clearly, the extra term inside the integral does not change the value of the cost. The function $p(\cdot)$ is reminiscent of the Lagrange multiplier function $\lambda(\cdot)$ in Section 2.5.2 (the exact relationship between the two will be clarified in Exercise 3.6 below). As we will see momentarily, p is also closely related to the momentum from Section 2.4. We will be working in the Hamiltonian framework, which is why we continue to use the same symbol p by which we denoted the momentum earlier (while some other sources prefer λ).

We will henceforth use the more explicit notation $\langle \cdot, \cdot \rangle$ for the inner product in $\mathbb{R}^n$. Let us introduce the *Hamiltonian*

$$H(t, x, u, p) := \langle p, f(t, x, u) \rangle - L(t, x, u). \quad (3.28)$$

Note that this definition matches our earlier definition of the Hamiltonian in calculus of variations, where we had $H(x, y, y', p) = \langle p, y' \rangle - L(x, y, y')$; we just need to remember that after we changed the notation from calculus of variations to optimal control, the independent variable x became t, the dependent variable y became x, its derivative y' became $\dot{x}$ and is given by (3.17), and the third argument of L is taken to be u rather than $\dot{x}$ (which with the current definition of H makes even more sense). We can rewrite the cost in terms of the Hamiltonian as

$$J(u) = \int_{t_0}^{t_1} \big(\langle p(t), \dot{x}(t) \rangle - H(t, x(t), u(t), p(t))\big) dt + K(x(t_1)). \quad (3.29)$$

3.4.2 First variation

We want to compute and analyze the first variation $\delta J|_{u^*}$ of the cost functional J at the optimal control function u^*. To do this, in view of the definition (1.32), we must isolate the first-order terms with respect to α in the cost difference between the perturbed control (3.23) and the optimal control u^*:

$$J(u) - J(u^*) = J(u^* + \alpha\xi) - J(u^*) = \delta J|_{u^*}(\xi)\alpha + o(\alpha). \qquad (3.30)$$

The formula (3.29) suggests to regard the difference $J(u) - J(u^*)$ as being composed of three distinct terms, which we now inspect in more detail. We will let $\approx$ denote equality up to terms of order $o(\alpha)$. For the terminal cost, we have

$$\begin{aligned} K(x(t_1)) - K(x^*(t_1)) &= K(x^*(t_1) + \alpha\eta(t_1) + o(\alpha)) - K(x^*(t_1)) \\ &\approx \langle K_x(x^*(t_1)), \alpha\eta(t_1)\rangle. \end{aligned} \qquad (3.31)$$

For the Hamiltonian, omitting the t-arguments inside x and u for brevity, we have

$$\begin{aligned} H(t,x,u,p) - H(t,x^*,u^*,p) &= H(t,x^* + \alpha\eta + o(\alpha), u^* + \alpha\xi, p) - \\ H(t,x^*,u^*,p) &\approx \langle H_x(t,x^*,u^*,p), \alpha\eta\rangle + \langle H_u(t,x^*,u^*,p), \alpha\xi\rangle. \end{aligned} \qquad (3.32)$$

As for the inner product $\langle p, \dot{x} - \dot{x}^*\rangle$, we use integration by parts as we did several times in calculus of variations:

$$\int_{t_0}^{t_1} \langle p(t), \dot{x}(t) - \dot{x}^*(t)\rangle dt = \langle p(t), x(t) - x^*(t)\rangle|_{t_0}^{t_1} - \int_{t_0}^{t_1} \langle \dot{p}(t), x(t) - x^*(t)\rangle dt$$

$$\approx \langle p(t_1), \alpha\eta(t_1)\rangle - \int_{t_0}^{t_1} \langle \dot{p}(t), \alpha\eta(t)\rangle dt \qquad (3.33)$$

where we used the fact that $x(t_0) = x^*(t_0)$. Combining the formulas (3.29)–(3.33), we readily see that the first variation is given by

$$\begin{aligned} \delta J|_{u^*}(\xi) = &- \int_{t_0}^{t_1} \big(\langle \dot{p} + H_x(t,x^*,u^*,p), \eta\rangle + \langle H_u(t,x^*,u^*,p), \xi\rangle \big) dt \\ &+ \langle K_x(x^*(t_1)) + p(t_1), \eta(t_1)\rangle \end{aligned} \qquad (3.34)$$

where η is related to ξ via the system (3.26).

The familiar first-order necessary condition for optimality (from Section 1.3.2) says that we must have $\delta J|_{u^*}(\xi) = 0$ for all ξ. This condition is true for every function p, but becomes particularly revealing if we make a special choice of p. Namely, let p be the solution of the differential equation

$$\dot{p} = -H_x(t,x^*,u^*,p) \qquad (3.35)$$

satisfying the boundary condition

$$p(t_1) = -K_x(x^*(t_1)). \tag{3.36}$$

Note that this boundary condition specifies the value of p at the *end* of the interval $[t_0, t_1]$, i.e., it is a final (or terminal) condition rather than an initial condition. In case of no terminal cost we treat K as being equal to 0, which corresponds to $p(t_1) = 0$. We label the function p defined by (3.35) and (3.36) as p^* from now on, to reflect the fact that it is associated with the optimal trajectory. We also extend the notation $|_*$ to mean evaluation along the optimal trajectory with $p = p^*$, so that, for example, $H|_*(t) = H(t, x^*(t), u^*(t), p^*(t))$. Setting $p = p^*$ and using the equations (3.35) and (3.36) to simplify the right-hand side of (3.34), we are left with

$$\delta J|_{u^*}(\xi) = -\int_{t_0}^{t_1} \langle H_u|_*, \xi \rangle \, dt = 0 \tag{3.37}$$

for all ξ. We know from Lemma 2.1[4] that this implies $H_u|_* \equiv 0$ or, in more detail,

$$H_u(t, x^*(t), u^*(t), p^*(t)) = 0 \qquad \forall\, t \in [t_0, t_1]. \tag{3.38}$$

The meaning of this condition is that the Hamiltonian has a stationary point as a function of u along the optimal trajectory. More precisely, the function $H(t, x^*(t), \cdot, p^*(t))$ has a stationary point at $u^*(t)$ for all t. This is just a reformulation of the property already discussed in Section 2.4.1 in the context of calculus of variations.

In light of the definition (3.28) of the Hamiltonian, we can rewrite our control system (3.17) more compactly as $\dot{x} = H_p(t, x, u)$. Thus the joint evolution of x^* and p^* is governed by the system

$$\begin{aligned} \dot{x}^* &= H_p|_*, \\ \dot{p}^* &= -H_x|_* \end{aligned} \tag{3.39}$$

which the reader will recognize as the system of Hamilton's canonical equations (2.30) from Section 2.4.1. Let us examine the differential equation for p^* in (3.39) in more detail. We can expand it with the help of (3.28) as

$$\dot{p}^* = -(f_x)^T\big|_* p^* + L_x|_*$$

where we recall that f_x is the Jacobian matrix of f with respect to x. This is a linear time-varying system of the form $\dot{p}^* = -A_*^T(t)p^* + L_x|_*$ where $A_*(\cdot)$ is the same as in the differential equation (3.27) derived earlier for the first-order state perturbation η. Two linear systems $\dot{x} = Ax$ and $\dot{z} = -A^T z$

---

[4]That lemma, translated to the present notation, requires $H_u|_*$ to be continuous (although it is clear from its proof that piecewise continuity is enough).

are said to be *adjoint* to each other, and for this reason p is called the *adjoint vector*; the system-theoretic significance of this concept will become clearer in Section 4.2.8. Note also that we can think of p as *acting* on the state or, more precisely, on the state velocity vector, since it always appears inside inner products such as $\langle p, \dot{x} \rangle$; for this reason, p is also called the *costate* (this notion will be further explored in Section 7.1).

The purpose of the next two exercises is to recover earlier conditions from calculus of variations, namely, the Euler-Lagrange equation and the Lagrange multiplier condition (for multiple degrees of freedom and several non-integral constraints) from the preliminary necessary conditions for optimality derived so far, expressed by the existence of an adjoint vector p^* satisfying (3.38) and (3.39).

Exercise 3.5   *The standard (unconstrained) calculus of variations problem with n degrees of freedom can be rewritten in the optimal control language by considering the control system*

$$\dot{x}_i = u_i, \qquad i = 1, \dots n$$

together with the cost $J(x) = \int_{t_0}^{t_f} L(t, x(t), \dot{x}(t)) dt$. Assuming that a given trajectory satisfies (3.38) and (3.39) for this system, prove that the Euler-Lagrange equations, $\frac{d}{dt} L_{\dot{x}_i} = L_{x_i}$, are satisfied along this trajectory.   □

Exercise 3.6   *Consider now a calculus of variations problem with n degrees of freedom and $k < n$ non-integral constraints, represented by the control system*

$$\dot{x}_i = f_i(t, x_1, \dots, x_n, u_1, \dots, u_{n-k}), \qquad\qquad i = 1, \dots, k,$$
$$\dot{x}_{k+i} = u_i, \qquad\qquad i = 1, \dots, n - k$$

with the same cost as in Exercise 3.5. Assuming that a given trajectory satisfies (3.38) and (3.39) for this system, prove that there exist functions $\lambda_i^* : [t_0, t_1] \to \mathbb{R}$, $i = 1, \dots, k$ such that the Euler-Lagrange equations for the augmented Lagrangian

$$L(t, x, \dot{x}) + \sum_{i=1}^{k} \lambda_i^*(t) \big(\dot{x}_i - f_i(t, x_1, \dots, x_n, \dot{x}_{k+1}, \dots, \dot{x}_n) \big)$$

are satisfied along this trajectory.   □

We caution the reader that the transition between the Hamiltonian and the Lagrangian requires some care, especially in the context of Exercise 3.6. The Lagrangian explicitly depends on $\dot{x}$ (and the partial derivatives $L_{\dot{x}_i}$ appear in the Euler-Lagrange equations), whereas the Hamiltonian should not (it should be a function of t, x, u, and p). The differential equations can be used to put H into the right form. A consequence of this, in Exercise 3.6, is that x will appear inside H in several different places. The Lagrange multipliers λ_i^* are related to the components of the adjoint vector p^*, but they are not the same.

3.4.3 Second variation

To better understand the behavior of H as a function of u along the optimal trajectory, let us bring in the second variation. To do this, we must first augment the description (3.24) of the perturbed state trajectory with the second-order term in α, writing $x = x^* + \alpha\eta + \alpha^2\zeta + o(\alpha^2)$. We then need to go back to the expressions (3.31)–(3.33) and expand them by adding second-order terms in α. With p already set equal to p^* as defined above, it is relatively straightforward to check that the ζ-dependent terms drop out—exactly in the same way as the η-dependent terms dropped out of the first variation (3.34)—and that the second variation is given by

$$\delta^2 J\big|_{u^*}(\xi) = -\frac{1}{2}\int_{t_0}^{t_1}\begin{pmatrix}\eta^T & \xi^T\end{pmatrix}\begin{pmatrix}H_{xx} & H_{xu} \\ H_{xu} & H_{uu}\end{pmatrix}\bigg|_* \begin{pmatrix}\eta \\ \xi\end{pmatrix}dt + \frac{1}{2}\eta^T K_{xx}(x^*(t_1))\eta$$

$$(3.40)$$

where we recall that η is obtained as the state of the system (3.26) driven by ξ.

Exercise 3.7 *Verify the formula (3.40).* □

We know from the second-order necessary condition for optimality (see Section 1.3.3) that we must have $\delta^2 J\big|_{u^*}(\xi) \geq 0$ for all ξ. Let us concentrate on the integrand in (3.40) and ask the following question: does one term in the Hessian matrix of H dominate the other terms? If yes, then this term should be nonpositive to ensure the correct sign of $\delta^2 J\big|_{u^*}(\xi)$. We encountered a very similar issue in Section 2.6.1 in relation to the inequality (2.61). We found there that for the overall integral to be nonnegative, the function multiplying $(\eta'(x))^2$ must be nonnegative, because η' may be large while η itself is small. The present situation is a bit different since ξ is not just the derivative of η, i.e., the system relating the two is not a simple integrator. However, the corresponding conclusion is still valid: $\xi^T H_{uu}\big|_* \xi$ is the dominant term because we may have a large ξ producing a small η (but not vice versa), as illustrated in Figure 3.5. Thus, in order for the second variation (3.40) to be nonnegative, we must have $\xi^T H_{uu}\big|_* \xi \leq 0$ for all ξ, which can only happen if the matrix $H_{uu}\big|_*$ is negative semidefinite:

$$H_{uu}(t, x^*(t), u^*(t), p^*(t)) \leq 0 \qquad \forall t \in [t_0, t_1]. \tag{3.41}$$

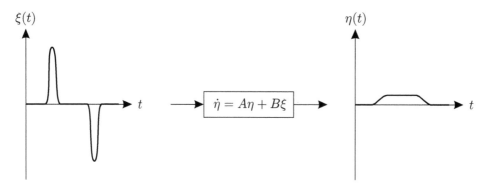

Figure 3.5: Large ξ can produce small η

This condition is known as the **Legendre-Clebsch condition** (in its Hamiltonian formulation).

We already know from (3.38) that for each $t \in [t_0, t_1]$, the function $H(t, x^*(t), \cdot, p^*(t))$ must have a stationary point at $u^*(t)$. The Legendre-Clebsch condition (3.41) tells us that if this stationary point is an extremum, then it is necessarily a *maximum*. Even though we have not proved that the stationary point must indeed be an extremum, it is tempting to conjecture that this Hamiltonian maximization property is true. In other words, our findings up to this point are very suggestive of the following (*not yet proved*) **necessary conditions for optimality**: *If $u^*(\cdot)$ is an optimal control and $x^*(\cdot)$ is the corresponding optimal state trajectory, then there exists an adjoint vector (costate) $p^*(\cdot)$ such that:*

1) *x^* and p^* satisfy the canonical equations (3.39) with respect to the Hamiltonian (3.28), with the boundary conditions $x^*(t_0) = x_0$ and $p^*(t_1) = -K_x(x^*(t_1))$.*

2) *For each fixed t, the function $u \mapsto H(t, x^*(t), u, p^*(t))$ has a (local) maximum at $u = u^*(t)$:*

$$H(t, x^*(t), u^*(t), p^*(t)) \geq H(t, x^*(t), u, p^*(t))$$

for all u near $u^*(t)$ and all $t \in [t_0, t_1]$.

3.4.4   Some comments

We remark that we could define the Hamiltonian and the adjoint vector using a different sign convention, as follows:

$$\widehat{H}(t, x, u, p) := \langle p, f(t, x, u) \rangle + L(t, x, u), \qquad \hat{p} := -p^*.$$

Then the function

$$\widehat{H}(t, x^*(t), \cdot, \hat{p}(t)) = -\langle p^*(t), f(t, x^*(t), \cdot)\rangle + L(t, x^*(t), \cdot)$$
$$= -H(t, x^*(t), \cdot, p^*(t))$$

would have a *minimum* at $u^*(t)$, while x^*, $\hat{p}$ would still satisfy the correct canonical equations with respect to $\widehat{H}$:

$$\dot{x}^* = f(t, x^*, u^*) = \widehat{H}_p(t, x^*, u^*)$$

and

$$\dot{\hat{p}} = -\dot{p}^* = H_x(t, x^*, u^*, p^*) = (f_x)^T\big|_* p^* - L_x\big|_*$$
$$= -(f_x)^T\big|_* \hat{p} - L_x\big|_* = -\widehat{H}_x(t, x^*, u^*, \hat{p}).$$

At first glance, this reformulation in terms of Hamiltonian minimization (rather than maximization) might seem more natural, because we are solving the minimization problem for the cost functional J. However, our problem is equivalent to the maximization problem for the functional $-J$ (defined by the running cost $-L$ and the terminal cost $-K$). So, whether we arrive at a *minimum principle* or a *maximum principle* is determined just by the sign convention, and has nothing to do with whether the cost functional is being minimized or maximized. There is no consensus in the literature on this choice of sign. The convention that we established in the previous subsections and will continue to follow in the rest of the book is consistent with our definition of the Hamiltonian in calculus of variations and its mechanical interpretation as the total energy of the system; see Sections 2.4.1 and 2.4.3. (In general, however, the cost in the optimal control problem is artificial from the physical point of view and is not related to Hamilton's action integral.)

Note that the necessary conditions for optimality from Section 3.4.3 are formulated as an existence statement for the adjoint vector p^* which arises directly as a solution of the second differential equation in (3.39); this is in contrast with Section 2.4.1 where the momentum p was first defined by the formula (2.28) and then a differential equation for it was obtained from the Euler-Lagrange equation. In the present setting, (3.38) and (3.39) encode all the necessary information about p^*, and we will find this approach more fruitful in optimal control. Observe that in the special case of the system $\dot{x} = u$ which corresponds to the unconstrained calculus of variations setting, we immediately obtain from (3.28) and (3.38) that p^* must be given by $p^* = L_u(t, x^*, u^*)$, and the momentum definition is recovered (up to the change of notation). This implies, in particular, that the Weierstrass necessary condition must also hold, in view of the calculation at the end of Section 3.1.2 (again modulo the change of notation).

The total derivative of the Hamiltonian with respect to time along an optimal trajectory is given by

$$\frac{d}{dt}\,H\big|_* = H_t\big|_* + \langle H_x\big|_*, \dot{x}^* \rangle + \langle H_p\big|_*, \dot{p}^* \rangle + \langle H_u\big|_*, \dot{u}^* \rangle = H_t\big|_* \qquad (3.42)$$

because the canonical equations (3.39) and the Hamiltonian stationarity condition (3.38) guarantee that the first two inner products cancel each other and the third one equals 0. In particular, if the problem is time-invariant in the sense that both f and L are independent of t, then $H_t = 0$ and we conclude that $H(x^*(t), u^*(t), p^*(t))$ must remain constant. If we want to think of H as the system's energy, the last statement says that this energy must be conserved along optimal trajectories.

We know that, at least in principle, we can obtain a second-order *sufficient* condition for optimality if we make appropriate assumptions to ensure that the second variation $\delta^2 J\big|_{u^*}$ is positive definite and dominates terms of order $o(\alpha^2)$ in $J(u^* + \alpha\xi) - J(u^*)$. While in general these assumptions take some work to write down and verify, the next exercise points to a case in which such a sufficient condition is easily established and applied (and the necessary condition becomes more tractable as well). This is the case when the system is linear and the cost is quadratic; we will study this class of problems in detail in Chapter 6.

Exercise 3.8   *For the fixed-time, free-endpoint optimal control problem studied in this section, assume in addition the following: the right-hand side of the control system takes the form $f(t, x, u) = A(t)x + B(t)u$ for arbitrary matrix functions A and B; the running cost takes the form $L(t, x, u) = x^T Q(t)x + u^T R(t)u$ where $Q(t)$ is symmetric positive semidefinite and $R(t)$ is symmetric positive definite for all t; and there is no terminal cost ($K \equiv 0$).*

a) Write down the canonical equations (3.39) for this case, and find a formula for the control satisfying the condition (3.38).

b) By analyzing the second variation, show that this control is indeed optimal (specify in what sense).   □

3.4.5 Critique of the variational approach and preview of the maximum principle

The variational approach presented in Sections 3.4.1–3.4.3 has led us, quite quickly, to the necessary conditions for optimality expressed by the canonical equations and the Hamiltonian maximization property (we did not actually prove the latter property, but we will see that it is indeed correct). While it helps us build intuition for what the correct statement of the maximum principle should look like, the variational approach has several limitations which, upon closer inspection, turn out to be quite severe.

CONTROL SET. Recall that our starting point was to consider perturbed controls of the form (3.23). Such perturbations make sense when the values of u^* are interior points of the control set U. This may not be the case, though, if U has a boundary, and bounded (or even finite) control sets are common in control applications. As we will see in the next chapter, the statement that the function $u \mapsto H(t, x^*(t), u, p^*(t))$ must have a maximum at $u^*(t)$ is true even in such situations and, moreover, this maximum is in fact global. However, we cannot hope to establish this fact using the variational approach, because $H_u|_*$ need not be 0 when the maximum is achieved at a boundary point of U.

FINAL STATE. In the preceding discussion we treated the case when the final state x_f is free, but we know (see Section 3.3.3) that in general we may have some target set S. Consider, for example, the case of a fixed endpoint: $S = \{t_1\} \times \{x_1\}$. Then the control perturbation ξ is no longer arbitrary, since the resulting state perturbation η must satisfy $\eta(t_1) = 0$. In view of the fact that η and ξ are related by the system (3.27), it is easy to show that admissible perturbations ξ must satisfy the constraint

$$\int_{t_0}^{t_1} \Phi_*(t_1, t) B(t) \xi(t) dt = 0$$

where $\Phi_*(\cdot, \cdot)$ is the state transition matrix for $A_*(\cdot)$ in (3.27). The second equation in (3.37) needs to hold only for admissible perturbations, and not for all ξ. This condition is no longer strong enough to let us conclude that $H_u|_* \equiv 0$. We see that the prospects of extending the variational approach beyond free-endpoint problems do not look very promising.

DIFFERENTIABILITY. When developing the first variation, we were tacitly assuming that H is differentiable with respect to u (as well as x). Since the Hamiltonian H is defined by (3.28), both f and L must thus be differentiable with respect to u. The reader can readily check that differentiability of f with respect to u was *not* one of the assumptions we made in Section 3.3.1 to ensure existence and uniqueness of solutions for our control system. In other words, the variational approach requires extra regularity assumptions to be imposed on the system. Having to assume differentiability of L with respect to u is also undesirable, as it rules out otherwise quite reasonable cost functionals like $J(u) = \int_{t_0}^{t_1} |u(t)| dt$. Furthermore, the analysis based on the second variation—which is needed to distinguish between a minimum and a maximum—involves second-order partial derivatives of H with respect to u. It is clear that the variational approach would take us on a path of overly restrictive regularity assumptions. Instead, we would like to establish the Hamiltonian maximization property more directly, not via derivatives.

CONTROL PERTURBATIONS. When considering the control and state perturbations as in (3.23) and (3.24) with α near 0, we are allowing only small deviations in both x and u. For the system $\dot{x} = u$, this would correspond exactly to the notion of a weak minimum from calculus of variations. However, as we already discussed as early as Section 2.2.1 (see in particular Example 2.1), we would like to have a larger family of control perturbations. More precisely, we want to capture optimality with respect to control perturbations that may be large, as long as the corresponding state trajectories are close to the given one. For example, Figure 3.6 illustrates a particular control perturbation (for controls that switch between only two values) which is very reasonable but falls outside the scope of the variational approach. As we will soon see, working with a richer perturbation family is crucial for obtaining sharper necessary conditions for optimality.

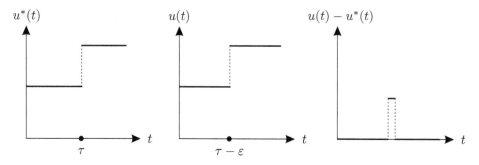

Figure 3.6: A control perturbation

In summary, while the basic form of the necessary conditions provided by the maximum principle will be similar to what we obtained using the variational approach, several shortcomings of the variational approach must be overcome in order to obtain a more satisfactory result. Specifically, we need to accommodate constraints on the control set, constraints on the final state, and weaker differentiability assumptions. A less restrictive notion of "closeness" of controls will be the key to achieving these goals. Borrowing a colorful expression from [PB94], we can describe the task ahead of us as "the cutting of the umbilical cord between the calculus of variations and optimal control theory." The maximum principle is a very nontrivial extension of the variational approach, and was developed many years later. The proof of the maximum principle is quite different from the argument given in this section; in particular, it is much more geometric in nature. We are now ready, in terms of both technical preparation and conceptual motivation, to tackle this proof in the next chapter.

3.5   NOTES AND REFERENCES FOR CHAPTER 3

Our derivation of the Weierstrass-Erdmann corner conditions is similar to the one given in [GF63, Section 15] (except that in [GF63] it is deduced from the general formula for the variation of a functional, while we present a self-contained argument). However, the discussion of these conditions in [GF63] does not correctly distinguish between weak and strong minima; see [Mac05, Chapter 11] for a more accurate treatment of this issue. The Weierstrass E-function is introduced in [GF63, Section 34] as well as [Mac05, Section 11.3]. A proof of the Weierstrass necessary condition is only sketched in [GF63]; a complete proof similar to ours can be found, e.g., in [Lei81] or [BM91]. The interested reader should also peruse McShane's paper [McS39], which is based on the monograph [Bli30] by Bliss. Both [Bli30] and [McS39] actually discuss more general constrained problems, and [Bli30] offers nice historical remarks at the end. For a careful look at the Hamiltonian reformulation of the Weierstrass necessary condition, see [Sus00, Handout 3]. A sufficient condition for a strong minimum based on the concept of a field is developed in [GF63, Sections 32–34].

The brachistochrone problem (Example 3.2 in Section 3.2) is studied in detail from both the calculus of variations and the optimal control viewpoints in the paper [SW97], which we already mentioned in the notes and references for Chapter 2. Conditions for existence and uniqueness of solutions, along the lines of Section 3.3.1, can be found in standard texts; see, e.g., [Kha02, Section 3.1] or [Son98, Appendix C.3] for ordinary differential equations without controls and [AF66, Section 3-18] or [Son98, Section 2.6] for control systems. The conditions given in [Son98] are sharper than those in [Kha02] and [AF66] but the proofs are essentially the same. Appendix C of [Son98] can also be consulted for more information on absolutely continuous and measurable functions.

From Section 3.3.2 until the end of the chapter, our primary reference was [AF66] (which, by the way, was apparently the first textbook on optimal control ever written). Cost functionals and transformations between their various forms are presented in [AF66, Sections 4-12, 5-14, 5-16]. The discussion on target sets is compiled from [AF66, Sections 4-12, 4-13, 5-12, 5-14]. Our presentation of the variational approach follows [AF66, Section 5-7], although the sign convention in [AF66] is the opposite of ours (see the explanation on page 97 above). Sufficient conditions for optimality in terms of the second variation—the subject that we only touched on in Exercise 3.8—are developed in [AF66, Section 5-8]. The material of Sections 3.4.4 and 3.4.5 is elaborated upon in [AF66, Sections 5-9, 5-10]. Finally, we mention that two-point boundary value problems of the kind that we saw in the necessary conditions for optimality stated in Section 3.4.3 (and will see again in the maximum principle) can be solved numerically using the shooting or forward-backward sweep methods; see, e.g., [DM70, Chapter 8] for detailed information on this topic.

Chapter Four

The Maximum Principle

4.1 STATEMENT OF THE MAXIMUM PRINCIPLE

This chapter is devoted to the maximum principle, which is in some sense the focal point of the book. We already previewed in the previous chapter the basic form of this result and the technical objectives that it is supposed to accomplish. We will now describe two special instances of the optimal control problem formulated in Section 3.3, for which we want to state and then (in Section 4.2) prove the maximum principle. Afterwards (in Section 4.3.1) we will explain how the maximum principle for other situations of interest can be deduced from these two specific cases.

4.1.1 Basic fixed-endpoint control problem

We will label as the *Basic Fixed-Endpoint Control Problem* the optimal control problem from Section 3.3 with the following additional specifications: $f = f(x, u)$ and $L = L(x, u)$, with no t-argument (the control system and the running cost are time-independent); f, f_x, L, and L_x are continuous (in other words, both f and L satisfy the stronger set of regularity conditions from Section 3.3.1); the target set is $S = [t_0, \infty) \times \{x_1\}$ (this is a free-time, fixed-endpoint problem); and $K \equiv 0$ (the terminal cost is absent). For this special problem, the maximum principle takes the following form.

Maximum Principle for the Basic Fixed-Endpoint Control Problem: *Let $u^* : [t_0, t_f] \to U$ be an optimal control (in the global sense) and let $x^* : [t_0, t_f] \to \mathbb{R}^n$ be the corresponding optimal state trajectory. Then there exist a function $p^* : [t_0, t_f] \to \mathbb{R}^n$ and a constant $p_0^* \leq 0$ satisfying $(p_0^*, p^*(t)) \neq (0, 0)$ for all $t \in [t_0, t_f]$ and having the following properties:*

1) *x^* and p^* satisfy the canonical equations*

$$
\begin{aligned}
\dot{x}^* &= H_p(x^*, u^*, p^*, p_0^*), \\
\dot{p}^* &= -H_x(x^*, u^*, p^*, p_0^*)
\end{aligned}
\tag{4.1}
$$

with the boundary conditions $x^*(t_0) = x_0$ *and* $x^*(t_f) = x_1$, *where the Hamiltonian* $H : \mathbb{R}^n \times U \times \mathbb{R}^n \times \mathbb{R} \to \mathbb{R}$ *is defined as*

$$H(x, u, p, p_0) := \langle p, f(x, u) \rangle + p_0 L(x, u). \tag{4.2}$$

2) *For each fixed* t, *the function* $u \mapsto H(x^*(t), u, p^*(t), p_0^*)$ *has a global maximum at* $u = u^*(t)$, *i.e., the inequality*

$$H(x^*(t), u^*(t), p^*(t), p_0^*) \geq H(x^*(t), u, p^*(t), p_0^*)$$

holds for all $t \in [t_0, t_f]$ *and all* $u \in U$.

3) $H(x^*(t), u^*(t), p^*(t), p_0^*) = 0$ *for all* $t \in [t_0, t_f]$.

A few clarifications are in order. First, the maximum principle, as stated here, describes necessary conditions for *global* optimality. However, we announced in Section 3.4.5 that one of our goals is to capture an appropriate notion of local optimality. The proof of the maximum principle will make it clear that the same conditions are indeed necessary for local optimality in the sense outlined in Section 3.4.5. We thus postpone further discussion of this issue until after the proof (see Section 4.3). Second, while the adjoint vector, or costate, p^* is already familiar from Section 3.4, one difference with the necessary conditions derived using the variational approach is the presence of p_0^*. This nonpositive scalar is called the *abnormal multiplier*. Similarly to the abnormal multiplier λ_0^* from Section 2.5, it equals 0 in degenerate cases; otherwise $p_0^* \neq 0$ and we can recover our earlier definition (3.28) of the Hamiltonian by normalizing $(p_0^*, p^*(t))$ so that $p_0^* = -1$ (note that such scaling does not affect any of the properties listed in the statement of the maximum principle). In the future, whenever the abnormal multiplier is not explicitly written, it is assumed to be equal to -1. Finally, in Section 3.4.4 we saw another scenario where H was constant, but the claim that $H \equiv 0$ may seem surprising. We will see later (in Section 4.3.1) that this is a special feature of free-time problems. The next exercise provides an early illustration of the usefulness of the above result.

Exercise 4.1   *Consider the optimal control formulation of the brachistochrone problem presented in Example 3.2 in Section 3.2. Use the maximum principle to confirm the fact (already known from Exercise 2.5) that optimal curves are cycloids given by (2.7).* $\square$

Let us now ask ourselves how restricted the Basic Fixed-Endpoint Control Problem really is. Time-independence of f and L and the absence of the terminal cost do not really introduce a loss of generality. Indeed, we know from Section 3.3 that we can eliminate t and K from the problem formulation by introducing the extra state variable $x_{n+1} := t$ (although this entails stronger regularity assumptions on the original right-hand side as a function of t) and passing to the new running cost $\hat{L} := L + K_t + K_x \cdot f$. On the other hand, the target set $S = [t_0, \infty) \times \{x_1\}$ is not very general, as it does not allow any flexibility in choosing the final state. This motivates us to consider the following refined problem formulation.

4.1.2   Basic variable-endpoint control problem

The *Basic Variable-Endpoint Control Problem* is the same as the Basic Fixed-Endpoint Control Problem except the target set is now of the form $S = [t_0, \infty) \times S_1$, where S_1 is a k-dimensional surface in $\mathbb{R}^n$ for some non-negative integer $k \leq n$. As in Section 1.2.2[1] we define such a surface via equality constraints:

$$S_1 = \{x \in \mathbb{R}^n : h_1(x) = h_2(x) = \cdots = h_{n-k}(x) = 0\}$$

where h_i, $i = 1, \ldots, n - k$ are $\mathcal{C}^1$ functions from $\mathbb{R}^n$ to $\mathbb{R}$. We also assume that every $x \in S_1$ is a regular point. As two extreme special cases, for $k = n$ we obtain $S_1 = \mathbb{R}^n$ (which gives a free-time, free-endpoint problem) while for $k = 0$ the surface S_1 reduces either to a single point (as in the Basic Fixed-Endpoint Control Problem) or to a set consisting of individual points. The difference between the maximum principle for this problem and the previous one lies only in the boundary conditions for the system of canonical equations.

Maximum Principle for the Basic Variable-Endpoint Control Problem: *Let $u^* : [t_0, t_f] \to U$ be an optimal control and let $x^* : [t_0, t_f] \to \mathbb{R}^n$ be the corresponding optimal state trajectory. Then there exist a function $p^* : [t_0, t_f] \to \mathbb{R}^n$ and a constant $p_0^* \leq 0$ satisfying $(p_0^*, p^*(t)) \neq (0,0)$ for all $t \in [t_0, t_f]$ and having the following properties:*

1) *x^* and p^* satisfy the canonical equations (4.1) with respect to the Hamiltonian (4.2), with the boundary conditions $x^*(t_0) = x_0$ and $x^*(t_f) \in S_1$.*

2) *$H(x^*(t), u^*(t), p^*(t), p_0^*) \geq H(x^*(t), u, p^*(t), p_0^*)$ for all $t \in [t_0, t_f]$ and all $u \in U$.*

3) *$H(x^*(t), u^*(t), p^*(t), p_0^*) = 0$ for all $t \in [t_0, t_f]$.*

4) *The vector $p^*(t_f)$ is orthogonal to the tangent space to S_1 at $x^*(t_f)$:*

$$\langle p^*(t_f), d \rangle = 0 \qquad \forall d \in T_{x^*(t_f)} S_1. \tag{4.3}$$

---

[1]The reader might need to revisit that section, as its terminology and notation will be freely used here.

The additional necessary condition (4.3) is called the *transversality condition* (we encountered its loose analog in Example 2.4 and Exercise 2.6 in Section 2.3.5). We know from Section 1.2.2 that the tangent space can be characterized as

$$T_{x^*(t_f)}S_1 = \{d \in \mathbb{R}^n : \langle \nabla h_i(x^*(t_f)), d \rangle = 0, \ i = 1, \dots, n - k\} \qquad (4.4)$$

and that (4.3) is equivalent to saying that $p^*(t_f)$ is a linear combination of the gradient vectors $\nabla h_i(x^*(t_f))$, $i = 1, \dots, n - k$. Note that when $k = n$ and hence $S_1 = \mathbb{R}^n$, the transversality condition reduces to $p^*(t_f) = 0$ (because the tangent space is the entire $\mathbb{R}^n$). On the other hand, the previous version of the maximum principle is a special case of the present result: when $S_1 = \{x_1\}$, its tangent space is 0 and (4.3) is true for all $p^*(t_f)$. In general, here as well as in the Basic Fixed-Endpoint Control Problem, we have n boundary conditions imposed on (x^*, p^*) at $t = t_0$ and n more at $t = t_f$. This gives the correct total number of boundary conditions to specify a solution of the $2n$-dimensional system (4.1). However, in the Basic Fixed-Endpoint Control Problem $x^*(t_f)$ was fixed and $p^*(t_f)$ was free, while here we have k degrees of freedom for $x^*(t_f) \in S_1$ but only $n - k$ degrees of freedom for $p^*(t_f) \perp T_{x^*(t_f)}S_1$. We see that the freer the state, the less free the costate: each additional degree of freedom for $x^*(t_f)$ eliminates one degree of freedom for $p^*(t_f)$.

The maximum principle was developed by the Pontryagin school in the Soviet Union in the late 1950s. It was presented to the wider research community at the first IFAC World Congress in Moscow in 1960 and in the celebrated book [PBGM62]. It is worth reflecting that the developments we have covered so far in this book—starting from the Euler-Lagrange equation, continuing to the Hamiltonian formulation, and culminating in the maximum principle—span more than 200 years. The progress made during this time period is quite remarkable, yet the origins of the maximum principle are clearly traceable to the early work in calculus of variations.

4.2   PROOF OF THE MAXIMUM PRINCIPLE

Our proof of the two versions of the maximum principle stated in the previous section will be divided into the following steps.

Step 1: From Lagrange to Mayer form As the first step, we will pass from the given Lagrange problem to an equivalent problem in the Mayer form by appending an additional state (this technique was already discussed in Section 3.3.2).

Step 2: Temporal control perturbation In the next step, we will apply a small perturbation to the length of the time interval over which the optimal control is acting, and will characterize the resulting perturbation of the terminal state.

Step 3: Spatial control perturbation In this step, we will replace the optimal control on a small time interval by an arbitrary constant control, and will study how the resulting state trajectory deviates from the optimal one at the end of this time interval. (This perturbation will be reminiscent of the one we used in the proof of the Weierstrass necessary condition in Section 3.1.2.)

Step 4: Variational equation We will then derive a linear differential equation which propagates, modulo terms of higher order, the effect of spatial control perturbations up to the terminal time. This is the variational equation (we already encountered this terminology in Section 2.6.2).

Step 5: Terminal cone Combining the effects of temporal and spatial control perturbations, we will construct a convex cone, with vertex at the terminal state of the optimal trajectory, which describes infinitesimal directions of all possible perturbations of the terminal state.

Step 6: Key topological lemma Next, we will use optimality to show that the terminal cone does not contain in its interior the direction of decreasing cost.

Step 7: Separating hyperplane We will invoke the separating hyperplane theorem to establish the existence of a hyperplane that passes through the terminal state and separates the terminal cone from the direction of decreasing cost. We will define the adjoint vector at the terminal time as the normal to this separating hyperplane.

Step 8: Adjoint equation We will then introduce the adjoint equation which propagates the adjoint vector up to the terminal time; it will match the second canonical equation, as required by the maximum principle.

Step 9: Properties of the Hamiltonian We will verify that the Hamiltonian maximization condition holds and that the Hamiltonian is identically 0 along the optimal trajectory. This will conclude the proof of the maximum principle for the Basic Fixed-Endpoint Control Problem.

Step 10: Transversality condition Finally, we will prove the maximum principle for the Basic Variable-Endpoint Control Problem by refining the separation property from steps 6 and 7 and arriving at the transversality condition.

Each of the subsections that follow corresponds to one step in the proof. Until we reach step 10 in Section 4.2.10, we assume that we are dealing with the Basic Fixed-Endpoint Control Problem, i.e., $S = [t_0, \infty) \times \{x_1\}$. The next exercise is meant to be worked on in parallel with following the proof.

Exercise 4.2   *Consider the double integrator* $\dot{x}_1 = x_2$, $\dot{x}_2 = u$ *with* $u \in [-1, 1]$. *Let the initial condition be* $x(0) = \begin{pmatrix} 0 \\ 0 \end{pmatrix}$, *let the final state be* $\begin{pmatrix} 2 \\ 2 \end{pmatrix}$, *and let the running cost be* $L \equiv 1$ *so that we have a time-optimal control problem. It is easy to check that the control* $u^*(t) = 1$, $t \in [0, 2]$ *is optimal. Specialize the main steps of the proof of the maximum principle to this problem and this control. Bring the relevant concepts and conclusions to the level of explicit numerical formulas, and draw figures wherever appropriate.* □

4.2.1   From Lagrange to Mayer form

As in Section 3.3.2, we define an additional state variable, $x^0 \in \mathbb{R}$, to be the solution of

$$\dot{x}^0 = L(x, u), \qquad x^0(t_0) = 0$$

and arrive at the augmented system

$$\begin{aligned} \dot{x}^0 &= L(x, u), \\ \dot{x} &= f(x, u) \end{aligned} \tag{4.5}$$

with the initial condition $\begin{pmatrix} 0 \\ x_0 \end{pmatrix}$. The cost can then be rewritten as

$$J(u) = \int_{t_0}^{t_f} \dot{x}^0(t)dt = x^0(t_f) \tag{4.6}$$

which means that in the new coordinates the problem is in the Mayer form (there is a terminal cost and no running cost). Also, the target set becomes $[t_0, \infty) \times \mathbb{R} \times \{x_1\} =: [t_0, \infty) \times S'$; here S' is the line in $\mathbb{R}^{n+1}$ that passes through $\begin{pmatrix} 0 \\ x_1 \end{pmatrix}$ and is parallel to the x^0-axis. To simplify the notation, we define

$$y := \begin{pmatrix} x^0 \\ x \end{pmatrix} \in \mathbb{R}^{n+1}$$

and write the system (4.5) more compactly as

$$\dot{y} = \begin{pmatrix} L(x, u) \\ f(x, u) \end{pmatrix} =: g(y, u) \tag{4.7}$$

(the right-hand side actually does not depend on x^0). Note that this system is well posed because we assumed that L has the same regularity properties as f.

An optimal trajectory $x^*(\cdot)$ of the original system in $\mathbb{R}^n$ corresponds in an obvious way to an optimal trajectory $y^*(\cdot)$ of the augmented system in $\mathbb{R}^{n+1}$. The first component $x^{0,*}$ of y^* describes the evolution of the cost in the original problem, and x^* is recovered from y^* by projection onto $\mathbb{R}^n$ parallel to the x^0-axis. This situation is depicted in Figure 4.1 (note that L is not necessarily positive, so x^0 need not actually be increasing along y^*). In this and all subsequent figures, the x^0-axis will be vertical.

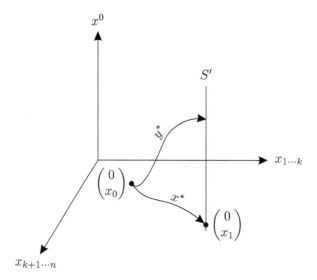

Figure 4.1: The optimal trajectory of the augmented system

From now on, we let t^* denote the terminal time of the optimal trajectory x^* (or, what is the same, of y^*). The next exercise offers a geometric interpretation of optimality; it will not be directly used in the current proof, but we will see a related idea in Section 4.2.6.

Exercise 4.3 *Let t_1 and t_2 be arbitrary time instants satisfying $t_0 \leq t_1 < t_2 \leq t^*$. Let S'' be the line passing through $\begin{pmatrix} 0 \\ x^*(t_2) \end{pmatrix}$ and parallel to the x^0- axis. Show that no trajectory y starting at $y^*(t_1) = \begin{pmatrix} x^{0,*}(t_1) \\ x^*(t_1) \end{pmatrix}$ can meet S'' below[2] the point $y^*(t_2) = \begin{pmatrix} x^{0,*}(t_2) \\ x^*(t_2) \end{pmatrix}$ at any time t_3, even not equal to t_2.* $\qquad\square$

In the particular case when $t_2 = t^*$, the claim in the exercise should be obvious: no other trajectory starting from some point $y^*(t_1)$ on the optimal trajectory can hit the line S' at a point lower than $y^*(t^*)$. In other words, a final portion of the optimal trajectory must itself be optimal with respect to its starting point as the initial condition. This idea, known as the *principle of optimality*, is illustrated in Figure 4.2. (The reader will notice that we are using different axes in different figures.)

---

[2]I.e., at a point with a smaller x^0-coordinate.

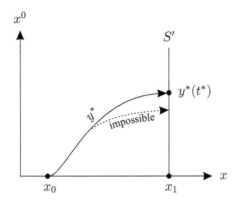

Figure 4.2: Principle of optimality

Another simple observation, which will be useful later, is that the Hamiltonian (4.2) can be equivalently represented as the following inner product in $\mathbb{R}^{n+1}$:

$$H(x, u, p, p_0) = \left\langle \begin{pmatrix} p_0 \\ p \end{pmatrix}, \begin{pmatrix} L(x, u) \\ f(x, u) \end{pmatrix} \right\rangle. \tag{4.8}$$

4.2.2   Temporal control perturbation

Let us see what happens if we introduce a small change in the terminal time t^* of the optimal trajectory, i.e., let the optimal control act on a little longer or a little shorter time interval. We formalize this as follows: for an arbitrary $\tau \in \mathbb{R}$ and a small $\varepsilon > 0$, we consider the perturbed control

$$u_\tau(t) := u^*(\min\{t, t^*\}), \qquad t \in [t_0, t^* + \varepsilon\tau]$$

which is illustrated by the thick curves in Figure 4.3 (for the two cases depending on the sign of τ).

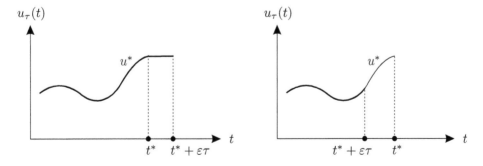

Figure 4.3: A temporal perturbation

We are interested in the value of the resulting perturbed trajectory y at the new terminal time $t^* + \varepsilon\tau$. For $\tau > 0$, the first-order Taylor expansion of y around $t = t^*$ gives

$$y(t^* + \varepsilon\tau) = y^*(t^*) + \dot{y}(t^*)\varepsilon\tau + o(\varepsilon) = y^*(t^*) + g(y^*(t^*), u^*(t^*))\varepsilon\tau + o(\varepsilon)$$
$$=: y^*(t^*) + \varepsilon\delta(\tau) + o(\varepsilon). \qquad (4.9)$$

For $\tau < 0$, we have $y(t^* + \varepsilon\tau) = y^*(t^* + \varepsilon\tau)$ and the first-order Taylor expansion of y^* around $t = t^*$ gives the same result. The vector $\varepsilon\delta(\tau)$ describes the infinitesimal (first-order in ε) perturbation of the terminal point. By definition, $\delta(\tau)$ depends linearly on τ. As we vary τ over $\mathbb{R}$, keeping ε fixed, the points $y^*(t^*) + \varepsilon\delta(\tau)$ form a line through $y^*(t^*)$. We denote this line by $\vec{\rho}$; see Figure 4.4. Every point on $\vec{\rho}$ corresponds to a control u_τ for some τ. On the other hand, the approximation of $y(t^* + \varepsilon\tau)$ by $y^*(t^*) + \varepsilon\delta(\tau)$ is valid only in the limit as $\varepsilon \to 0$. So, $\delta(\tau)$ tells us the direction—but not the magnitude—of the terminal point deviation caused by an infinitesimal change in the terminal time. The arrow over ρ is meant to indicate that points on the line correspond to perturbation *directions*. Note that we are describing deviations of the terminal point in the (x^0, x)-space only, ignoring the differences in the terminal times; accordingly, the time axis is not included in the figures.

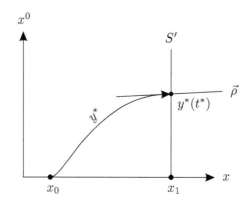

Figure 4.4: The effect of a temporal control perturbation

4.2.3   Spatial control perturbation

We now construct control perturbations known as "needle" perturbations, or Pontryagin-McShane perturbations. As the former name suggests, they will be represented by pulses of short duration; the reason for the latter name is that perturbations of this kind were first used by McShane in calculus of variations (see Section 3.1.2) and later adopted by Pontryagin's school for the proof of the maximum principle.

Let w be an arbitrary element of the control set U. Consider the interval $I := (b - \varepsilon a, b] \subset (t_0, t^*)$, where $b \neq t^*$ is a point of continuity[3] of u^*, $a > 0$ is arbitrary, and $\varepsilon > 0$ is small. We define the perturbed control

$$u_{w,I}(t) := \begin{cases} u^*(t) & \text{if } t \notin I, \\ w & \text{if } t \in I. \end{cases}$$

Figure 4.5 illustrates this control perturbation and the resulting state trajectory perturbation. As the figure suggests, the perturbed trajectory y corresponding to $u_{w,I}$ will deviate from y^* on the interval I and afterwards will "run parallel" to y^*. We now proceed to formally characterize the deviation over I; the behavior of y over the interval $[b, t^*]$ will be studied in Section 4.2.4.

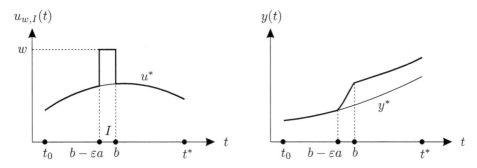

Figure 4.5: A spatial control perturbation and its effect on the trajectory

We will let $\approx$ denote equality up to terms of order $o(\varepsilon)$. The first-order Taylor expansion of y^* around $t = b$ gives

$$y^*(b - \varepsilon a) \approx y^*(b) - \dot{y}^*(b)\varepsilon a. \tag{4.10}$$

Rearranging terms and using the fact that y^* satisfies the differential equation (4.7) with $u = u^*$, we have

$$y^*(b) \approx y^*(b - \varepsilon a) + g(y^*(b), u^*(b))\varepsilon a. \tag{4.11}$$

On the other hand, the first-order Taylor expansion of the perturbed solution y around $t = b - \varepsilon a$ yields

$$y(b) \approx y(b - \varepsilon a) + \dot{y}(b - \varepsilon a)\varepsilon a$$

where by $\dot{y}(b - \varepsilon a)$ we mean the right-sided derivative of y at $t = b - \varepsilon a$. Since $y(b - \varepsilon a) = y^*(b - \varepsilon a)$ by construction and y satisfies (4.7) with $u = u_{w,I}$, we obtain

$$y(b) \approx y^*(b - \varepsilon a) + g(y^*(b - \varepsilon a), w)\varepsilon a. \tag{4.12}$$

---

[3]The reason for this assumption is that the subsequent Taylor expansions rely on y being differentiable at $t = b$.

We now apply the Taylor expansion to the last term in (4.12):

$$g(y^*(b-\varepsilon a), w)\varepsilon a \approx g(y^*(b), w)\varepsilon a + g_y(y^*(b), w)(y^*(b-\varepsilon a) - y^*(b))\varepsilon a. \quad (4.13)$$

In view of (4.10), the second term on the right-hand side of (4.13) is of order ε^2; hence we can omit it and the approximation will remain valid. Thus (4.12) simplifies to

$$y(b) \approx y^*(b - \varepsilon a) + g(y^*(b), w)\varepsilon a.$$

Comparing this formula with (4.11), we arrive at

$$y(b) \approx y^*(b) + \nu_b(w)\varepsilon a \qquad\qquad (4.14)$$

where

$$\nu_b(w) := g(y^*(b), w) - g(y^*(b), u^*(b)). \qquad\qquad (4.15)$$

Intuitively, this result makes sense: up to terms of order $o(\varepsilon)$, the difference between the two states $y(b)$ and $y^*(b)$ is the difference (4.15) between the state velocities at $y = y^*(b)$ corresponding to $u = w$ and $u = u^*(b)$, multiplied by the length εa of the time interval on which the perturbation is acting.

4.2.4   Variational equation

We are now interested in how the difference between the trajectory y arising from a spatial (needle) perturbation and the optimal trajectory y^* propagates after the perturbation stops acting, i.e., for $t \geq b$. To study this question, let us begin by writing

$$y(t) = y^*(t) + \varepsilon\psi(t) + o(\varepsilon) =: y(t, \varepsilon) \qquad\qquad (4.16)$$

for $b \leq t \leq t^*$, where $\psi : [b, t^*] \to \mathbb{R}^{n+1}$ is a quantity that we want to characterize and ε is the same as in Section 4.2.3. We know from (4.14) that $\psi(b)$ exists and is given by

$$\psi(b) = \nu_b(w)a. \qquad\qquad (4.17)$$

We now derive a differential equation that ψ must satisfy on the time interval $(b, t^*]$, where both y and y^* have the same corresponding control u^*. It is clear from (4.16) that

$$\psi(t) = y_\varepsilon(t, 0) \qquad\qquad (4.18)$$

(the existence of this partial derivative, and therefore of ψ, for $t > b$ will become evident in a moment). Let us rewrite the system (4.7) as an integral equation:

$$y(t, \varepsilon) = y(b, \varepsilon) + \int_b^t g(y(s, \varepsilon), u^*(s))ds.$$

Differentiating both sides of this equation with respect to ε at $\varepsilon = 0$ and using (4.16) with $t = b$ and (4.17), we obtain

$$y_\varepsilon(t,0) = \nu_b(w)a + \int_b^t g_y(y(s,0), u^*(s))y_\varepsilon(s,0)ds$$

which, in view of (4.16) and (4.18), amounts to

$$\psi(t) = \nu_b(w)a + \int_b^t g_y(y^*(s), u^*(s))\psi(s)ds.$$

Taking the derivative with respect to t, we conclude that ψ satisfies the differential equation

$$\dot{\psi} = g_y(y^*, u^*)\psi = g_y|_* \psi. \tag{4.19}$$

We will use this equation to describe how spatial perturbations propagate with time. Pictorially, the role of ψ is illustrated in Figure 4.6, with the understanding that the labels involving ψ are accurate only up to terms of order $o(\varepsilon)$.

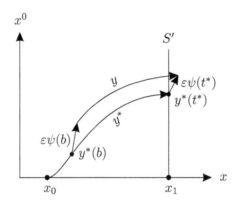

Figure 4.6: Propagation of a spatial perturbation

The equation (4.19) can be written in the form

$$\dot{\psi} = A_*(t)\psi \tag{4.20}$$

where $A_*(t) := g_y|_* (t)$. This system is simply the linearization of the original system (4.7) around the optimal trajectory y^*. Recall that a more detailed description of the system (4.7) is given by (4.5). Letting (η^0, η) be the corresponding components of ψ and writing out the variational equation (4.19) in terms of these components, we easily arrive at

$$\dot{\eta}^0 = (L_x)^T|_* \eta,$$
$$\dot{\eta} = f_x|_* \eta$$

(because L and f do not depend on x^0). Here $(L_x)^T$ is a row vector (the transpose of the gradient of L with respect to x) and f_x is an $n \times n$ matrix (the Jacobian matrix of f with respect to x). The resulting more explicit formula for A_* is

$$A_*(t) = \begin{pmatrix} 0 & (L_x)^T|_*(t) \\ 0 & f_x|_*(t) \end{pmatrix}. \tag{4.21}$$

The equation (4.19) is in fact the *variational equation* corresponding to the system (4.7), according to the terminology introduced in Section 2.6.2. The equation (3.26) in Section 3.4 is also similar and was similarly derived, although it served a different purpose (it described the effect of a small control perturbation on a given trajectory, while here we are studying the difference between two nearby trajectories with the same control). Also, we see that ε in the definition of a needle perturbation essentially corresponds to α in the variational approach; we chose to use different symbols for these two parameters because of the different specific ways in which they are introduced.

The value of ψ at the terminal time t^* gives us an approximation of the terminal point $y(t^*)$ of the perturbed trajectory. Namely, from (4.16) evaluated at $t = t^*$ we have

$$y(t^*) = y^*(t^*) + \varepsilon\psi(t^*) + o(\varepsilon). \tag{4.22}$$

Let us denote by $\Phi_*(\cdot, \cdot)$ the state transition matrix for the linear time-varying system (4.20), so that

$$\psi(t^*) = \Phi_*(t^*, b)\psi(b).$$

The initial value $\psi(b)$ is given by (4.17), hence

$$\psi(t^*) = \Phi_*(t^*, b)\nu_b(w)a$$

where $\nu_b(w)$ was defined in (4.15). Plugging this expression into (4.22), we obtain

$$y(t^*) = y^*(t^*) + \varepsilon\Phi_*(t^*, b)\nu_b(w)a + o(\varepsilon). \tag{4.23}$$

Let us introduce the notation

$$\delta(w, I) := \Phi_*(t^*, b)\nu_b(w)a \tag{4.24}$$

to arrive at the more compact formula

$$y(t^*) = y^*(t^*) + \varepsilon\delta(w, I) + o(\varepsilon)$$

(here the interval I used for constructing the needle perturbation encodes information about the values of a and b).

4.2.5   Terminal cone

We now want to describe geometrically the combined effect of the temporal and spatial control perturbations on the terminal state. The vector $\varepsilon\delta(w, I)$ describes the infinitesimal (first-order in ε) perturbation of the terminal state caused by the needle perturbation with parameters w and I. (It corresponds to the vector labeled as $\varepsilon\psi(t^*)$ in Figure 4.6.) From the definition (4.24) of $\delta(w, I)$, it is clear that its *direction* depends only on w and b, but not on a. We let $\vec{\rho}(w, b)$ denote the ray in this direction originating at $y^*(t^*)$. If we keep w, b, and ε fixed, $\vec{\rho}(w, b)$ consists of the points $y^*(t^*) + \varepsilon\delta(w, I)$ for various values of a. The construction of $\vec{\rho}(w, b)$ is analogous to that of $\vec{\rho}$ in Section 4.2.2, except that $\vec{\rho}(w, b)$ is unidirectional (because both a and ε are positive) whereas $\vec{\rho}$ was bidirectional. We also let $\vec{P}$ denote the union of the rays $\vec{\rho}(w, b)$ for all possible values of w and b. Then $\vec{P}$ is a cone with vertex at $y^*(t^*)$. Note that this cone is not convex; for example, if the control set U (in which w takes values) is finite, then $\vec{P}$ will in general be a union of isolated rays starting at $y^*(t^*)$.

Let us now ask ourselves the following question: is there a spatial control perturbation such that the corresponding first-order perturbation of the terminal point is, say, $\varepsilon\delta(w_1, I_1) + \varepsilon\delta(w_2, I_2)$ for some control values w_1, w_2 and intervals $I_1 = (b_1 - \varepsilon a_1, b_1]$ and $I_2 = (b_2 - \varepsilon a_2, b_2]$? We will now see that the right way to "add" two needle perturbations is to concatenate them, i.e., to perturb u^* *both* on I_1 (by setting it equal to w_1 there) and on I_2 (by setting it equal to w_2). Here we are assuming that $b_1 < b_2$, so that for ε small enough I_1 and I_2 do not overlap. Such a spatial perturbation is shown in Figure 4.7.

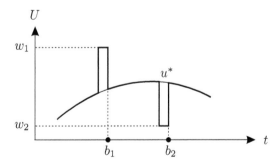

Figure 4.7: "Adding" spatial perturbations

The resulting first-order perturbation of the terminal point will then be the sum $\varepsilon\delta(w_1, I_1) + \varepsilon\delta(w_2, I_2)$. This is true simply because the variational equation, which propagates first-order state perturbations up to the terminal time, is linear. Indeed, according to the formulas derived in the two previous subsections, we will have

$$y(b_1) = y^*(b_1) + \nu_{b_1}(w_1)\varepsilon a_1 + o(\varepsilon)$$

at the end of the first perturbation interval, then

$$y(b_2) = y^*(b_2) + \varepsilon\big(\Phi_*(b_2, b_1)\nu_{b_1}(w_1)a_1 + \nu_{b_2}(w_2)a_2\big) + o(\varepsilon)$$

at the end of the second perturbation interval, and finally, by the semigroup property $\Phi_*(t^*, b_2)\Phi_*(b_2, b_1) = \Phi_*(t^*, b_1)$ of the state transition matrix,

$$
\begin{aligned}
y(t^*) &= y^*(t^*) + \varepsilon\Phi_*(t^*, b_2)\big(\Phi_*(b_2, b_1)\nu_{b_1}(w_1)a_1 + \nu_{b_2}(w_2)a_2\big) + o(\varepsilon) \\
&= y^*(t^*) + \varepsilon\Phi_*(t^*, b_1)\nu_{b_1}(w_1)a_1 + \varepsilon\Phi_*(t^*, b_2)\nu_{b_2}(w_2)a_2 + o(\varepsilon) \\
&= y^*(t^*) + \varepsilon\delta(w_1, I_1) + \varepsilon\delta(w_2, I_2) + o(\varepsilon).
\end{aligned}
$$

Exercise 4.4   *Explain how to "add" two needle perturbations with $b_1 = b_2$. In other words, given two control values w_1 and w_2 and two intervals $I_1 = (b - \varepsilon a_1, b]$ and $I_2 = (b - \varepsilon a_2, b]$ with the same right endpoint, construct a spatial control perturbation such that the resulting terminal point satisfies $y(t^*) = y^*(t^*) + \varepsilon\delta(w_1, I_1) + \varepsilon\delta(w_2, I_2) + o(\varepsilon)$. (Keep in mind that $w_1 + w_2$ might not be an admissible control value.)*   □

More generally, if we want to generate the infinitesimal perturbation $\varepsilon\beta_1\delta(w_1, I_1) + \varepsilon\beta_2\delta(w_2, I_2)$ for some $\beta_1, \beta_2 > 0$, then it is easy to see that we need to adjust the lengths of the intervals on which the two needle perturbations are acting: we need to set $u(t) = w_1$ on $\bar{I}_1 := (b_1 - \varepsilon\beta_1 a_1, b_1]$ and $u(t) = w_2$ on $\bar{I}_2 := (b_2 - \varepsilon\beta_2 a_2, b_2]$. It is also clear that this construction immediately extends to linear combinations (with positive coefficients) of more than two terms.

Recall that $\vec{P}$ is the cone with vertex at $y^*(t^*)$ formed by the rays $\vec{\rho}(w, b)$ corresponding to all simple (individual) needle perturbations. The preceding discussion demonstrates that by concatenating different needle perturbations, we generate a larger cone (with the same vertex) which consists exactly of convex combinations of points in $\vec{P}$. We denote this convex cone by $\mathrm{co}(\vec{P})$.

In Section 4.2.2 we also constructed the line $\vec{\rho}$ of perturbation directions arising from the temporal perturbations of u^*. We now add this line to the convex cone $\mathrm{co}(\vec{P})$, in the sense of adding vectors attached to the point $y^*(t^*)$. More precisely, we consider the set of points of the form

$$y = y^*(t^*) + \varepsilon\Big(\beta_0\delta(\tau) + \sum_{i=1}^{m}\beta_i\delta(w_i, I_i)\Big) \tag{4.25}$$

where $\varepsilon > 0$, $\beta_0, \beta_1, \ldots, \beta_m \geq 0$, $\delta(\tau)$ comes from (4.9) for some τ, and each $\delta(w_i, I_i)$ comes from (4.24) for some w_i and I_i. We denote this set by C_{t^*}

and call it the *terminal cone*. It is easy to check that C_{t^*} is again a convex cone, with vertex at $y^*(t^*)$. This construction is illustrated in Figure 4.8, where C_{t^*} is the infinite "wedge" between the two half-planes.

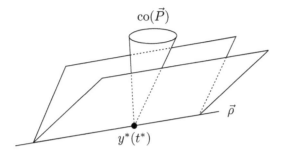

Figure 4.8: The terminal cone

By the same reasoning as before, we can show that for every point $y \in C_{t^*}$ given by (4.25) there exists a perturbation of u^* such that the terminal point of the perturbed trajectory satisfies

$$y(t_f) = y + o(\varepsilon).$$

To obtain the desired control perturbation, we need to apply a concatenated spatial perturbation as explained above, followed by the temporal perturbation that adjusts the terminal time by $\beta_0 \varepsilon \tau$. Since the intervals I_i are strictly inside $[t_0, t^*]$, they do not interfere with the temporal perturbation (for small enough ε). The fact that the resulting total first-order perturbation of the terminal point is indeed the correct one hinges on the linearity of the variational equation and on the linear dependence of $\delta(\tau)$ on τ.

4.2.6 Key topological lemma

Up until now, we have not yet used the fact that u^* is an optimal control and y^* is an optimal trajectory. As discussed in Section 4.2.1 and demonstrated in Figure 4.2, optimality means that no other trajectory y corresponding to another control u can reach the line S' (the vertical line through $\begin{pmatrix} 0 \\ x_1 \end{pmatrix}$ in the y-space) at a point below $y^*(t^*)$. Since the terminal cone C_{t^*} is a linear approximation of the set of points that we can reach by applying perturbed controls, we expect that the terminal cone should face "upward."

To formalize this observation, consider the vector

$$\mu := \begin{pmatrix} -1 & 0 & \cdots & 0 \end{pmatrix}^T \in \mathbb{R}^{n+1} \tag{4.26}$$

and let $\vec{\mu}$ be the ray generated by this vector (which points downward) originating at $y^*(t^*)$. Optimality suggests that $\vec{\mu}$ should be directed outside of C_{t^*}, a situation illustrated in Figure 4.9. Since C_{t^*} is only an approximation, the correct claim is actually slightly weaker.

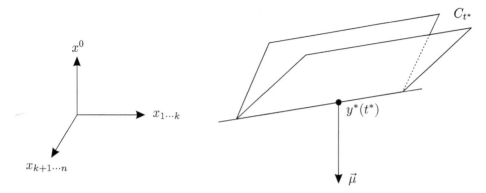

Figure 4.9: Illustrating the statement of Lemma 4.1

LEMMA 4.1. $\vec{\mu}$ *does not intersect the interior of the cone* C_{t^*}.

In other words, $\vec{\mu}$ can in principle touch C_{t^*} along the boundary, but it cannot lie inside it. We note that since C_{t^*} is a cone, $\vec{\mu}$ intersects its interior if and only if all points of $\vec{\mu}$ except $y^*(t^*)$ are interior points of C_{t^*}.

Let us see what would happen if the statement of the lemma were false and $\vec{\mu}$ were inside C_{t^*}. By construction of the terminal cone, as explained at the end of Section 4.2.5, there would exist a (spatial plus temporal) perturbation of u^* such that the terminal point of the perturbed trajectory y would be given by

$$y(t_f) = y^*(t^*) + \varepsilon\beta\mu + o(\varepsilon)$$

for some (arbitrary) $\beta > 0$. Writing this out in terms of the components (x^0, x) of y and recalling the definition (4.26) of μ and the relation (4.6) between x^0 and the cost, we obtain

$$J(u) = J(u^*) - \varepsilon\beta + o(\varepsilon),$$
$$x(t_f) = x_1 + o(\varepsilon)$$

where u is the perturbed control that generates y. Presently there is no direct contradiction with optimality of u^* yet, because the terminal point $x(t_f)$ of the perturbed trajectory x is different from the prescribed terminal point x_1, i.e., x need not hit the target set. Thus we see that although Lemma 4.1 certainly seems plausible, it is not obvious.

Let us try to build a more convincing argument in support of Lemma 4.1. If we suppose that the statement of the lemma is false, then we can pick a point $\hat{y}$ on the ray $\vec{\mu}$ below $y^*(t^*)$ such that $\hat{y}$ is contained in C_{t^*} together with a ball of some positive radius ε around it; let us denote this ball by B_ε. For a suitable value of $\beta > 0$, we have $\hat{y} = y^*(t^*) + \varepsilon\beta\mu$. Since the points in B_ε belong to C_{t^*}, they are of the form (4.25) and can be written as

$y^*(t^*)+\varepsilon\nu$ where the vectors $\varepsilon\nu$ are first-order perturbations of the terminal point arising from control perturbations constructed earlier. We know that the actual terminal points of trajectories corresponding to these control perturbations are given by

$$y^*(t^*) + \varepsilon\nu + o(\varepsilon). \qquad (4.27)$$

We denote the set of these terminal points by $\widetilde{B}_\varepsilon$; we can think of it as a "warped" version of B_ε, since it is $o(\varepsilon)$ away from B_ε.

In the above discussion, $\varepsilon > 0$ was fixed; we now make it tend to 0. The point $y^*(t^*) + \varepsilon\beta\mu$, which we relabel as $\hat{y}_\varepsilon$ to emphasize its dependence on ε, will approach $y^*(t^*)$ along the ray $\vec{\mu}$ as $\varepsilon \to 0$ (here β is the same fixed positive number as in the original expression for $\hat{y}$). The ball B_ε, which now stands for the ball of radius ε around $\hat{y}_\varepsilon$, will still belong to C_{t^*} and consist of the points $y^*(t^*) + \varepsilon\nu$ for each value of ε. Terminal points of perturbed state trajectories (the perturbations being parameterized by ε) will still generate a "warped ball" $\widetilde{B}_\varepsilon$ consisting of points of the form (4.27). Figure 4.10 should help visualize this construction.

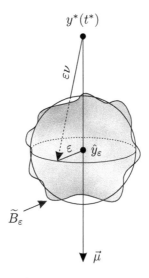

Figure 4.10: Proving Lemma 4.1

Since the center of B_ε is on $\vec{\mu}$ below $y^*(t^*)$, the radius of B_ε is ε, and the "warping" is of order $o(\varepsilon)$, for sufficiently small ε the set $\widetilde{B}_\varepsilon$ will still intersect the ray $\vec{\mu}$ below $y^*(t^*)$. But this means that there exists a perturbed trajectory x which hits the desired terminal point x_1 with a lower value of the cost. The resulting contradiction proves the lemma.

The above claim about a nonempty intersection of $\widetilde{B}_\varepsilon$ and $\vec{\mu}$ seems intuitively obvious. The original proof of the maximum principle in [PBGM62] states that this fact is obvious, but then adds a lengthy footnote explaining that a rigorous proof can be given using topological arguments. A conceivable scenario that must be ruled out is one in which the set $\widetilde{B}_\varepsilon$ has a hole (or dent) in it and the ray $\vec{\mu}$ goes through this hole. It turns out that this is indeed impossible, thanks to continuity of the "warping" map that transforms B_ε to $\widetilde{B}_\varepsilon$. In fact, it can be shown that $\widetilde{B}_\varepsilon$ *contains*, for ε small enough, a ball centered at $\hat{y}_\varepsilon$ whose radius is of order $\varepsilon - o(\varepsilon)$. One quick way to prove this is by applying Brouwer's fixed point theorem (which states that a continuous map from a ball to itself must have a fixed point).

Exercise 4.5 *Prove rigorously that $\widetilde{B}_\varepsilon$ and $\vec{\mu}$ must have a nonempty intersection for sufficiently small ε.* □

4.2.7 Separating hyperplane

A standard result in convex analysis known as the Separating Hyperplane Theorem (see, e.g., [Ber99, Proposition B.13] or [BV04, Section 2.5.1]) says that if C and D are two nonempty disjoint convex sets then there exists a hyperplane that *separates* them; by this we mean that C is contained in one of the two closed half-spaces created by the hyperplane and D is contained in the other. The ray $\vec{\mu}$ is a convex set, and from the convexity of the terminal cone C_{t^*} it is easy to see that its interior is convex as well. Lemma 4.1 guarantees that $\vec{\mu}$ does not intersect the interior of C_{t^*}. Therefore, we can apply the Separating Hyperplane Theorem to conclude the existence of a hyperplane separating $\vec{\mu}$ from the interior of C_{t^*}, and hence from C_{t^*} itself.[4] Obviously, this separating hyperplane must pass through the point $y^*(t^*)$ which is a common point of C_{t^*} and $\vec{\mu}$. Such a hyperplane need not be unique. The normal to the hyperplane is a nonzero vector in $\mathbb{R}^{n+1}$ (it is defined up to a constant multiple once we fix the hyperplane). Let us denote this normal vector by

$$\begin{pmatrix} p_0^* \\ p^*(t^*) \end{pmatrix} \tag{4.28}$$

where $p_0^* \in \mathbb{R}$ and $p^*(t^*) \in \mathbb{R}^n$ are, by definition, its x^0-component and x-component, respectively. Then the equation of the hyperplane is

$$\left\langle \begin{pmatrix} p_0^* \\ p^*(t^*) \end{pmatrix}, y \right\rangle = \left\langle \begin{pmatrix} p_0^* \\ p^*(t^*) \end{pmatrix}, y^*(t^*) \right\rangle$$

and the separation property is formally written as

$$\left\langle \begin{pmatrix} p_0^* \\ p^*(t^*) \end{pmatrix}, \delta \right\rangle \leq 0 \qquad \forall \delta \text{ such that } y^*(t^*) + \delta \in C_{t^*} \tag{4.29}$$

---

[4]There is one degenerate case in which we cannot directly apply the Separating Hyperplane Theorem as described, namely, the case when the interior of C_{t^*} is empty. However, it can be shown that if the interior of the convex set C_{t^*} is empty then there exists a hyperplane that contains C_{t^*} (see, e.g., [dlF00, p. 238]), and this hyperplane trivially separates $\vec{\mu}$ and C_{t^*}.

and

$$\left\langle \begin{pmatrix} p_0^* \\ p^*(t^*) \end{pmatrix}, \mu \right\rangle \geq 0 \qquad (4.30)$$

where μ is the vector (4.26) which generates the ray $\vec{\mu}$. Note that if we were to flip the direction of the normal vector (4.28), the inequality signs in (4.29) and (4.30) would be reversed; the present choice is simply a matter of sign convention (cf. Section **3.4.4**). For an illustration, see Figure 4.11 in which the shaded object represents the separating hyperplane.

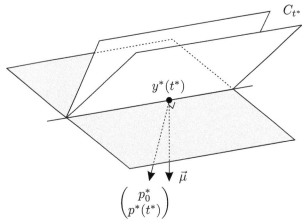

Figure 4.11: A separating hyperplane for the Basic Fixed-Endpoint Control Problem

In view of the definition (4.26) of μ, the inequality (4.30) simply says that $p_0^* \leq 0$, as required by the statement of the maximum principle. This will be the only use of (4.30). The normal vector (4.28) will serve as the terminal condition for the adjoint system, to be defined next.

4.2.8   Adjoint equation

As we already mentioned on page 94, two (time-varying) linear systems of the form $\dot{x} = Ax$ and $\dot{z} = -A^T z$ are called *adjoint* to each other. Solutions of adjoint systems are linked by the property that their inner product $\langle z(t), x(t) \rangle$ remains constant, as shown by the following simple calculation:

$$\frac{d}{dt}\langle z, x \rangle = \langle \dot{z}, x \rangle + \langle z, \dot{x} \rangle = (-A^T z)^T x + z^T A x = 0.$$

We now consider a specific pair of adjoint systems on the time interval $[t_0, t^*]$. As the first system (the x-system in the preceding discussion) we take the variational equation (4.19), described in more detail by the equations (4.20) and (4.21). The second, adjoint system is then

$$\dot{z} = -A_*^T(t)z = \begin{pmatrix} 0 & 0 \\ -\left. L_x \right|_*(t) & -\left. (f_x)^T \right|_*(t) \end{pmatrix} z \qquad (4.31)$$

where in the last expression, obtained from (4.21), L_x is a column vector (it is the gradient of L with respect to x) and $(f_x)^T$ is the transpose of the Jacobian matrix of f with respect to x. Let us denote the first component of z by p_0 and the vector of the remaining n components of z by p. Then the first differential equation in (4.31) reads $\dot{p}_0 = 0$ while the rest of the system (4.31) becomes

$$\dot{p} = - L_x|_* p_0 - (f_x)^T|_* p$$

which, in view of the definition (4.2) of the Hamiltonian, is equivalent to

$$\dot{p} = -H_x(x^*, u^*, p, p_0).$$

Now, let us specify the terminal condition for the system (4.31) at time t^* by setting $z(t^*)$ equal to the vector (4.28). Further, we relabel the p-component of the solution z corresponding to this terminal condition as p^*. This gives $p_0(t) = p_0^*$ for all t and

$$\dot{p}^* = -H_x(x^*, u^*, p^*, p_0^*)$$

which is the second canonical equation in (4.1). With a slight abuse of terminology, we will sometimes refer to this differential equation as the adjoint equation. It is easy to see that the first canonical equation in (4.1) also holds by the definition of H; thus statement 1 of the maximum principle has been established. By the aforementioned property of adjoint systems, we have

$$\left\langle \begin{pmatrix} p_0^* \\ p^*(t) \end{pmatrix}, \psi(t) \right\rangle = \left\langle \begin{pmatrix} p_0^* \\ p^*(t^*) \end{pmatrix}, \psi(t^*) \right\rangle \qquad \forall\, t \in [t_0, t^*] \qquad (4.32)$$

for every solution ψ of the variational equation (4.19).

The vector (4.28), which is normal to the separating hyperplane, is nonzero. Since (4.31) is a homogeneous (unforced) linear time-varying system, we have

$$\begin{pmatrix} p_0^* \\ p^*(t) \end{pmatrix} \neq 0 \qquad \forall\, t \in [t_0, t^*] \qquad (4.33)$$

as required in the statement of the maximum principle. Geometrically, we can think of the vector in (4.33) as the normal vector to a hyperplane passing through $y^*(t)$. We can then associate the solution of the adjoint system to a family of hyperplanes that is "flowing back" along the optimal trajectory. In view of (4.32), the perturbed trajectory associated with ψ always remains on the same side of the hyperplane.

4.2.9 Properties of the Hamiltonian

We are now ready to prove the remaining properties of the Hamiltonian, namely, statements 2 and 3 of the maximum principle for the Basic Fixed-Endpoint Control Problem (see Section 4.1.1).

STATEMENT 2: HAMILTONIAN MAXIMIZATION CONDITION

Let us go back to the formula (4.23), which says that the infinitesimal perturbation of the terminal point caused by a needle perturbation of the optimal control with parameters w, b, a is described by the vector $\varepsilon\Phi_*(t^*, b)\nu_b(w)a$. We subsequently labeled this vector as $\varepsilon\delta(w, I)$, and by construction $y^*(t^*) + \varepsilon\delta(w, I)$ belongs to the terminal cone C_{t^*}. Applying the inequality (4.29) which encodes the separating hyperplane property, and noting that ε and a are both positive, we have

$$\left\langle \begin{pmatrix} p_0^* \\ p^*(t^*) \end{pmatrix}, \Phi_*(t^*, b)\nu_b(w) \right\rangle \leq 0.$$

Next, since Φ_* is the state transition matrix for the variational equation (4.19), we know that $\Phi_*(t^*, b)\nu_b(w)$ is the value at time t^* of the solution of the variational equation passing through $\nu_b(w)$ at time b. Invoking the adjoint property (4.32), we obtain

$$\left\langle \begin{pmatrix} p_0^* \\ p^*(b) \end{pmatrix}, \nu_b(w) \right\rangle \leq 0. \tag{4.34}$$

Since $\nu_b(w)$ was defined in (4.15) and $g(y, u)$ was defined in (4.7), we have

$$\nu_b(w) = g(y^*(b), w) - g(y^*(b), u^*(b)) = \begin{pmatrix} L(x^*(b), w) - L(x^*(b), u^*(b)) \\ f(x^*(b), w) - f(x^*(b), u^*(b)) \end{pmatrix}.$$

We can thus expand (4.34) as follows:

$$\left\langle \begin{pmatrix} p_0^* \\ p^*(b) \end{pmatrix}, \begin{pmatrix} L(x^*(b), w) \\ f(x^*(b), w) \end{pmatrix} \right\rangle \leq \left\langle \begin{pmatrix} p_0^* \\ p^*(b) \end{pmatrix}, \begin{pmatrix} L(x^*(b), u^*(b)) \\ f(x^*(b), u^*(b)) \end{pmatrix} \right\rangle.$$

Recalling the expression (4.8) for the Hamiltonian, we see that this is equivalent to

$$H(x^*(b), w, p^*(b), p_0^*) \leq H(x^*(b), u^*(b), p^*(b), p_0^*). \tag{4.35}$$

In the above derivation, w was an arbitrary element of the control set U and b was an arbitrary time in the interval (t_0, t^*) at which the optimal control u^* is continuous. Thus we have established that the Hamiltonian maximization condition holds everywhere except possibly a finite number of time instants (discontinuities of u^*). Additionally, recall that u^* is piecewise continuous and we adopted the convention (see page 84) that the value of a piecewise continuous function at each discontinuity is equal to the limit either from the left or from the right. Letting b in (4.35) approach a discontinuity of u^* or an endpoint of $[t_0, t^*]$ from an appropriate side, and using continuity of x^* and p^* in time and continuity of H in all variables, we see that the Hamiltonian maximization condition must actually hold everywhere.

This conclusion can be understood intuitively as follows. The Hamiltonian is the inner product of the augmented adjoint vector $\begin{pmatrix} p_0^* \\ p^* \end{pmatrix}$ with the right-hand side of the augmented control system (the velocity of y). When the optimal control is perturbed, the state trajectory deviates from the optimal one in a direction that makes a nonpositive inner product with the augmented adjoint vector (at the time when the perturbation stops acting). Therefore, such control perturbations can only decrease the Hamiltonian, regardless of the value of the perturbed control during the perturbation interval.

STATEMENT 3: $H|_* \equiv 0$

The separation property (4.29) applies, in particular, to $\delta = \delta(\tau) \in C_{t^*}$, the terminal state perturbation vector corresponding to a temporal perturbation of the control. We know from (4.9) and (4.7) that this vector is given by

$$\delta(\tau) = \begin{pmatrix} L(x^*(t^*), u^*(t^*)) \\ f(x^*(t^*), u^*(t^*)) \end{pmatrix} \tau.$$

Since τ can be either positive or negative, the inequality (4.29) can be satisfied only if

$$\left\langle \begin{pmatrix} p_0^* \\ p^*(t^*) \end{pmatrix}, \begin{pmatrix} L(x^*(t^*), u^*(t^*)) \\ f(x^*(t^*), u^*(t^*)) \end{pmatrix} \right\rangle = 0.$$

By virtue of (4.8), this is equivalent to

$$H(x^*(t^*), u^*(t^*), p^*(t^*), p_0^*) = 0.$$

In other words, $H|_*$ equals 0 at the terminal time.[5] We need to prove that it is 0 everywhere.

Let us show that $H|_*(\cdot) = H(x^*(\cdot), u^*(\cdot), p^*(\cdot), p_0^*)$ is a continuous function of time, even though the optimal control u^* need not be continuous. The argument that follows is very similar to the one that the reader presumably used a while ago to solve Exercise 3.3 on page 80. Let t be a point of discontinuity of u^*. Of course, x^* and p^* are continuous everywhere. Applying the Hamiltonian maximization condition (4.35) with $b < t$ and $w = u^*(t^+)$ and making b approach t from the left, we have

$$H(x^*(t), u^*(t^+), p^*(t), p_0^*) \le H(x^*(t), u^*(t^-), p^*(t), p_0^*).$$

Similarly, applying (4.35) with $b > t$ and $w = u^*(t^-)$, in the limit as b approaches t from the right we obtain

$$H(x^*(t), u^*(t^+), p^*(t), p_0^*) \ge H(x^*(t), u^*(t^-), p^*(t), p_0^*).$$

---

[5] Here and below, when using the notation $|_*$ we also mean that $p_0 = p_0^*$.

Thus the two quantities must actually be equal, and the continuity claim is established.

Next, let us show that the function $H|_*$ is constant. In Section 3.4.4, in the context of the variational approach, we established this property by simply differentiating the Hamiltonian with respect to time, but here we need to be more careful because the existence of H_u has not been assumed. In view of the Hamiltonian maximization condition, we can write

$$H(x^*(t), u^*(t), p^*(t), p_0^*) = m(x^*(t), p^*(t))$$

where

$$m(x, p) := \max_{u \in U} H(x, u, p, p_0^*).$$

We just saw that $m(x^*(\cdot), p^*(\cdot))$ is a continuous function of time. For an arbitrary pair of times t, t', we have the inequalities

$$\begin{aligned}
H(x^*(t'), u^*(t), p^*(t'), p_0^*) &- H(x^*(t), u^*(t), p^*(t), p_0^*) \\
&\leq m(x^*(t'), p^*(t')) - m(x^*(t), p^*(t)) \\
&\leq H(x^*(t'), u^*(t'), p^*(t'), p_0^*) - H(x^*(t), u^*(t'), p^*(t), p_0^*).
\end{aligned} \tag{4.36}$$

In view of the standing assumptions made at the beginning of Section 4.1.1 and the canonical equations (4.1), it is straightforward to show that the function $H(x^*(\cdot), u^*(\bar{t}), p^*(\cdot), p_0^*)$ is continuously differentiable for each fixed $\bar{t} \in [t_0, t^*]$, with an upper bound on the magnitude of its derivative independent of $\bar{t}$. From this fact and (4.36) we easily conclude that the function $m(x^*(\cdot), p^*(\cdot))$ is locally Lipschitz. Therefore, it is absolutely continuous, and hence differentiable for almost all t (see page 85). We can now study its derivative.

Exercise 4.6 *Show that $m(x^*(\cdot), p^*(\cdot))$ has zero derivative (wherever the derivative exists). Do this by considering, for a pair of nearby times t, t', the limit*

$$\lim_{t' \to t} \frac{m(x^*(t'), p^*(t')) - m(x^*(t), p^*(t))}{t' - t}$$

and using the inequalities (4.36).   □

We have shown that the function $H(x^*(\cdot), u^*(\cdot), p^*(\cdot), p_0^*)$ equals 0 at $t = t^*$, is continuous everywhere, and has zero derivative almost everywhere. Thus it is identically 0, as claimed. Our proof of the maximum principle for the Basic Fixed-Endpoint Control Problem is now complete. At this point the reader should also finish Exercise 4.2.

4.2.10   Transversality condition

We now turn to the Basic Variable-Endpoint Control Problem. In this case there is an additional statement to be proved, which is the transversality condition (4.3). In Section 4.2.6 we had that the terminal cone C_{t^*} was separated from the ray $\vec{\mu}$; the reason for this was that hitting a point on $\vec{\mu}$ below $y^*(t^*)$ contradicted optimality. When the fixed endpoint x_1 is replaced by the surface S_1, we would instead have a contradiction with optimality if we were able to hit a point with a cost lower than $x^{0,*}(t^*)$ whose x-component is in S_1 (but is not necessarily $x^*(t^*)$). Let us denote the set of such points by D. We are looking to establish separation between convex sets; for this reason, just as we replace the actual set of terminal points with its linear approximation C_{t^*}, we will consider the linear approximation of D given by the linear span of $\vec{\mu}$ and the tangent space $T_{x^*(t^*)}S_1$, i.e., the set

$$ T := \left\{ y \in \mathbb{R}^{n+1} : y = y^*(t^*) + \begin{pmatrix} 0 \\ d \end{pmatrix} + \beta\mu, \ d \in T_{x^*(t^*)}S_1, \ \beta \geq 0 \right\}. \quad (4.37) $$

In Figure 4.12, T is the shaded "semi-plane." The subspace $T_{x^*(t^*)}S_1$, when translated to $y^*(t^*)$, gives the upper line in the figure. D is the surface (more precisely, the open semi-surface) consisting of points which lie "directly under" the upper curve in the figure. T is tangent to D along $\vec{\mu}$; both T and D are bounded from above but extend infinitely far down (and T includes its upper boundary while D does not).

LEMMA 4.2. *T does not intersect the interior of the cone C_{t^*}.*

This lemma is proved by an appropriate generalization of the argument we used to prove Lemma 4.1. Suppose that the statement is not true. Then we can find a point in T which is contained in C_{t^*} together with some ε-ball B_ε around it. We can write this point as

$$ \hat{y}_\varepsilon = y^*(t^*) + \varepsilon \begin{pmatrix} 0 \\ d \end{pmatrix} + \varepsilon\beta\mu $$

for suitable $d \in T_{x^*(t^*)}S_1$ and $\beta > 0$ (moving $\hat{y}_\varepsilon$ slightly down if necessary, we can ensure that it does not lie on the upper boundary of T). Since B_ε belongs to C_{t^*}, each point in B_ε is given by $y^*(t^*)+\varepsilon\nu$ where $\varepsilon\nu$ is a first-order terminal state perturbation arising from a temporal and/or spatial control perturbation. We know that the corresponding exact terminal states are of the form $y^*(t^*) + \varepsilon\nu + o(\varepsilon)$ and form a "warped" version of B_ε, which we denote by $\widetilde{B}_\varepsilon$.

This construction remains valid as $\varepsilon \to 0$. The ball B_ε is centered at $\hat{y}_\varepsilon \in T$, its radius is ε, and the "warping" that produces $\widetilde{B}_\varepsilon$ is of order $o(\varepsilon)$. As in Exercise 4.5, it can be shown that $\widetilde{B}_\varepsilon$ contains a ball centered

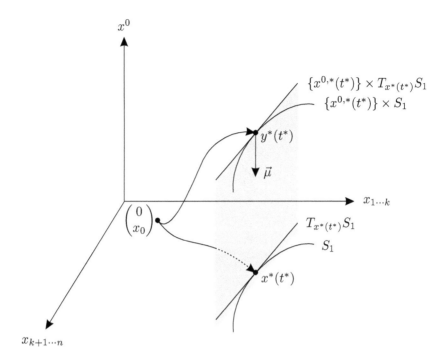

Figure 4.12: Illustrating the construction of the set T

at $\hat{y}_\varepsilon$ whose radius is of order $\varepsilon - o(\varepsilon)$. Furthermore, since T and D are tangent to each other along $\vec{\mu}$, the distance from $\hat{y}_\varepsilon$ to D is also of order $o(\varepsilon)$. Hence, for ε small enough, $\widetilde{B}_\varepsilon$ actually intersects D. But this, as we already noted, contradicts optimality of y^*, and the lemma is established. The preceding argument is illustrated in Figure 4.13, where the plane and the curved surface represent T and D, respectively, the shaded object is the portion of $\widetilde{B}_\varepsilon$ that lies between T and D, and the ray in T containing $\hat{y}_\varepsilon$, $\varepsilon > 0$ is also shown.

By Lemma 4.2 and the Separating Hyperplane Theorem, there exists a hyperplane that separates T and C_{t^*}. We denote its normal vector by (4.28) as before. Figure 4.14 depicts C_{t^*}, T, and the separating hyperplane. In view of the definition (4.37) of T and the fact that $d = 0$ belongs to $T_{x^*(t^*)}S_1$, the separation property still gives us the inequalities (4.29) and (4.30). Thus all the constructions and conclusions of Sections 4.2.8 and 4.2.9 still apply, and so we know that the first three statements of the maximum principle are true. On the other hand, writing the separation property for vectors in T with β (the $\vec{\mu}$-component) equal to 0, we obtain the additional inequality

$$\left\langle \begin{pmatrix} p_0^* \\ p^*(t^*) \end{pmatrix}, \begin{pmatrix} 0 \\ d \end{pmatrix} \right\rangle = \langle p^*(t^*), d \rangle \geq 0 \qquad \forall\, d \in T_{x^*(t^*)}S_1. \tag{4.38}$$

For each $d \in T_{x^*(t^*)}S_1$ we also have $-d \in T_{x^*(t^*)}S_1$, as is clear from (4.4). This fact and (4.38) imply that actually $\langle p^*(t^*), d \rangle = 0$ for all $d \in T_{x^*(t^*)}S_1$,

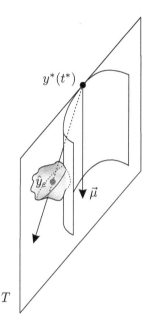

Figure 4.13: Proving Lemma 4.2

which is precisely the desired transversality condition (4.3). Our proof of the maximum principle for the Basic Variable-Endpoint Control Problem is now complete.

Note that in the special case when $S_1 = \mathbb{R}^n$ (a free-time, free-endpoint problem), the hyperplane separates C_{t^*} from the entire $(n+1)$-dimensional half-space that lies below $y^*(t^*)$. Clearly, this hyperplane must be horizontal, hence its normal must be vertical and we conclude that $p^*(t^*) = 0$. This is consistent with (4.3) because $T_{x^*(t^*)}S_1 = \mathbb{R}^n$ in this case.

4.3 DISCUSSION OF THE MAXIMUM PRINCIPLE

Our main objective in the remainder of this chapter is to gain a better understanding of the maximum principle by discussing and interpreting its statement and by applying it to specific classes of problems. We begin this task here by making a few technical remarks.

One should always remember that the maximum principle provides *necessary* conditions for optimality. Thus it only helps single out optimal control candidates, each of which needs to be further analyzed to determine whether it is indeed optimal. The reader should also keep in mind that an optimal control may not even exist (the existence issue will be addressed in detail in Section 4.5). For many problems of interest, however, the optimal solution does exist and the conditions provided by the maximum principle

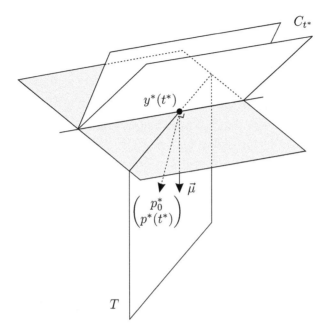

Figure 4.14: A separating hyperplane for the Basic Variable-Endpoint Control Problem

are strong enough to help identify it, either directly or after a routine additional elimination process. We already saw an example supporting this claim in Exercise 4.1 and will study other important examples in Section 4.4.

When stating the maximum principle, we ignored the distinction between different kinds of local minima by working with a globally optimal control u^*, i.e., by assuming that $J(u^*) \leq J(u)$ for all other admissible controls u that produce state trajectories satisfying the given endpoint constraint. However, it is clear from the proof that global optimality was not used. The control perturbations used in the proof produced controls u which differ from u^* on a small interval of order ε in length, making the $\mathcal{L}_1$ norm of the difference, $\int_{t_0}^{t_f} |u(t) - u^*(t)| dt$, small for small ε. The resulting perturbed trajectory x, on the other hand, was close to the optimal trajectory x^* in the sense of the 0-norm, i.e., $\max_{t_0 \leq t \leq t_f} |x(t) - x^*(t)|$ was small for small ε (as is clear from the calculations given in Sections 4.2.2–4.2.4). It can be shown that the conditions of the maximum principle are in fact necessary for local optimality when closeness in the (x, u)-space is measured by the 0-norm for x and $\mathcal{L}_1$ norm for u; we stress that the Hamiltonian maximization condition (statement 2 of the maximum principle) remains global. At this point it may be instructive to think of the system $\dot{x} = u$ as an example and to recall the discussion in Section 3.4.5 related to Figure 3.6. In that context, the notion of a local minimum with respect to the norm we just described is in between the notions of weak and strong minima; indeed, weak minima are defined with respect to the 0-norm for both x and u, while strong minima are with respect to the 0-norm for x with no constraints on u. For strong minima, the necessary conditions provided by the maximum principle are still valid. This is not the case for weak minima, because in a needle perturbation the control value w is no longer arbitrary: it must be close to $u^*(b)$.

The statement of the maximum principle contains the condition (justified in Section 4.2.8) that $(p_0^*, p^*(t)) \neq (0,0)$ for all t. In fact, since the origin in $\mathbb{R}^{n+1}$ is an equilibrium of the linear adjoint system (4.31), if p_0^*, $p^*(t)$ vanish for some t then they must vanish for all t. Thus, the above condition could be equivalently stated as $(p_0^*, p^*(t)) \neq (0,0)$ for *some* t. This condition is sometimes called the *nontriviality condition*, because with $(p_0^*, p^*) \equiv (0,0)$ all the statements of the maximum principle are trivially satisfied. In some cases, it is possible to show that the adjoint vector itself, $p^*(t)$, is nonzero for all t. For example, suppose that the running cost L is everywhere nonzero (this is true, for instance, in time-optimal control problems, where $L \equiv 1$). The Hamiltonian satisfies $H|_* = \langle p^*, f|_* \rangle + p_0^* L|_* \equiv 0$ (by statement 3 of the maximum principle). If $p^*(t) = 0$ for some t, then we have $p_0^* L|_* (t) = 0$, hence $p_0^* = 0$ and we reach a contradiction with the nontriviality condition. We will give another example later involving a terminal cost (see Exercise 4.7 below). As for the abnormal multiplier p_0^*, since it is the vertical coordinate of the normal to the separating hyperplane, $p_0^* = 0$ corresponds to the case when the separating hyperplane is vertical (and cannot be tilted). The projection of such a hyperplane onto the x-space is a hyperplane in $\mathbb{R}^n$, and all perturbed controls must bring the state x to the same side of this projected hyperplane. In a majority of control problems this does not happen and we can set $p_0^* = -1$. We also know that the separating hyperplane cannot be vertical and p_0^* cannot be 0 in the free-endpoint case (see the end of Section 4.2.10).

4.3.1   Changes of variables

Some control problems that we are interested in do not fit into the setting of the Basic Fixed-Endpoint or Variable-Endpoint Control Problem. We now want to discuss several such scenarios and, for each one, to arrive at the correct statement of the maximum principle. One way to do this is to formally reduce a problem at hand to one of the two basic problems considered above by changing variables; the changes of variables that we will use were already discussed in Section 3.3. Another approach is to adapt the proof of the maximum principle to these new situations; this task can be somewhat more challenging but also more insightful as it helps us better understand the proof. These various cases also help clarify why we say "maximum *principle*" (rather than calling it a theorem): this terminology reflects the fact that specific versions of the result for different problems are different, but share the same basic features.

FIXED TERMINAL TIME

Suppose that the terminal time t_f is fixed, so that the terminal time t^* of the optimal trajectory x^* considered in the proof must equal a given value t_1. Temporal control perturbations are then no longer admissible. Accordingly, the line $\vec{\rho}$ formed by the perturbation directions $\delta(\tau)$ defined in (4.9) must not be used when generating the terminal cone C_{t^*}. We now invite the reader to check exactly where these perturbation directions $\delta(\tau)$ were used in the proof of the maximum principle. As a matter of fact, they were used in one place only: to show that $H|_*(t^*) = 0$. Thus we conclude that the Hamiltonian will now be constant but not necessarily 0 along the optimal trajectory, while all the other conditions remain unchanged.

Let us confirm this fact by reducing the fixed-time problem to a free-time one via the familiar trick of introducing the extra state variable $x_{n+1} := t$. The system becomes

$$\begin{aligned} \dot{x} &= f(x, u), & x(t_0) &= x_0, \\ \dot{x}_{n+1} &= 1, & x_{n+1}(t_0) &= t_0. \end{aligned} \tag{4.39}$$

If the original target set was $\{t_1\} \times S_1$, then for the new system we can write the target set as $[t_0, \infty) \times S_1 \times \{t_1\}$, with the terminal time no longer explicitly constrained. The maximum principle for the Basic Variable-Endpoint Control Problem can now be applied. The Hamiltonian for the new problem is $\overline{H} = \langle p, f \rangle + p_{n+1} + p_0 L = H + p_{n+1}$, where $H = \langle p, f \rangle + p_0 L$ is the Hamiltonian for the original fixed-time problem. Clearly, the differential equation for p^* is the same as before and the Hamiltonian maximization condition for $\overline{H}$ implies the one for H. We know that $\overline{H}|_* = H|_* + p^*_{n+1} \equiv 0$. Moreover, we have $\dot{p}^*_{n+1} = -\overline{H}_{x_{n+1}}|_* = -H_t|_* = 0$ (since H does not depend on t) hence p^*_{n+1} is constant. Thus $H|_*$ is indeed equal to a constant, namely, $-p^*_{n+1}$. It is no longer guaranteed to be 0; in fact, since the final value of x_{n+1} is fixed, the transversality condition gives us no information about p^*_{n+1}. This property of the Hamiltonian is also consistent with what we saw in Section 3.4.4.

TIME-DEPENDENT SYSTEM AND COST

The same idea of appending the state variable $x_{n+1} = t$ and passing to the system (4.39) can be applied when the original system's right-hand side f and/or the running cost L depend on t. The Hamiltonian is now time-dependent:

$$H(t, x, u, p, p_0) := \langle p, f(t, x, u) \rangle + p_0 L(t, x, u). \tag{4.40}$$

The previous discussion remains valid up to and including the equation $\dot{p}^*_{n+1} = -H_t|_*$ but the right-hand side no longer equals 0. Thus, p^*_{n+1} and $H|_* = -p^*_{n+1}$ are not constant any more. Instead, we have the differential equation

$$\frac{d}{dt} H|_* = H_t|_* \tag{4.41}$$

with the boundary condition $H|_* (t_f) = -p^*_{n+1}(t_f)$. If the terminal time t_f in the original problem is free, then the final value of x_{n+1} is free and the transversality condition yields $p^*_{n+1}(t_f) = 0$. In this case we obtain $H|_* (t_f) = 0$ and, integrating (4.41), $H|_* (t) = -\int_t^{t_f} H_t|_* (s)ds$. Note that (4.41) is consistent with the equation obtained in (3.42) in the context of the variational approach (although the middle portion of (3.42) does not apply here).

The same conclusion can be reached by following the proof of the maximum principle and verifying that it carries over to the time-varying scenario without major changes, except that the argument showing that $H|_*$ is constant becomes invalid and only (4.41) can be established (the reader who solved Exercise 4.6 will readily see why). We can appreciate, however, that in the present case the method of changing the variables is much simpler and more reliable.

TERMINAL COST

Let us now consider a situation where a terminal cost of the form $K = K(x_f)$ is present. To illustrate just one simple case, we suppose that we are dealing with a free-time, free-endpoint problem in the Mayer form, i.e., there is no running cost ($L \equiv 0$). We assume the function K to be differentiable as many times as desired; everything else is as in the Basic Variable-Endpoint Control Problem.

Since $L \equiv 0$, there is no need to consider the x^0-coordinate. Temporal and spatial control perturbations can be used to construct the terminal cone C_{t^*} as before, except that it now lives in the original x-space: $C_{t^*} \subset \mathbb{R}^n$. By optimality, and since the terminal state is free, no perturbed state trajectory can have a terminal cost lower than $K(x^*(t^*))$, the terminal cost of the candidate optimal trajectory. Thus we expect that K should not decrease along any direction in C_{t^*}, a property that we can write as

$$\langle -K_x(x^*(t^*)), \delta \rangle \leq 0 \qquad \forall \delta \text{ such that } x^*(t^*) + \delta \in C_{t^*}. \tag{4.42}$$

Geometrically, this means that C_{t^*} lies on one side of the hyperplane passing through $x^*(t^*)$ with normal $-K_x(x^*(t^*))$, which we henceforth assume to be a nonzero vector. A comparison of (4.42) with (4.29) unmistakably suggests that we should define

$$p^*(t^*) := -K_x(x^*(t^*)). \tag{4.43}$$

Indeed, we know that in a free-endpoint problem the final value of the costate should be completely constrained (to give the correct total number of boundary conditions for the system of canonical equations). With the above definition of $p^*(t^*)$ the inequality (4.42) matches (4.29), following which the adjoint equation and the analysis of the Hamiltonian

$$H(x, u, p) = \langle p, f(x, u) \rangle \tag{4.44}$$

are developed in the same way as before. The companion inequality (4.30) in our earlier proof of the maximum principle was only needed to show that $p_0^* \leq 0$; here we do not have such an inequality, and we do not need it because there is no p_0^*.

The above argument is of course not rigorous, so let us again validate our conjecture by formally reducing the present scenario to the Basic Variable-Endpoint Control Problem. A transformation that accomplishes this goal is provided by the formula

$$K(x_f) = K(x_0) + \int_{t_0}^{t_f} \langle K_x(x(t)), f(x(t), u(t)) \rangle \, dt.$$

Ignoring $K(x_0)$ which is a known constant, we arrive at an equivalent problem in the Lagrange form with $L = \langle K_x, f \rangle$. For this modified problem, the Hamiltonian is $\overline{H} = \langle \bar{p}, f \rangle + p_0 \langle K_x, f \rangle = \langle \bar{p} + p_0 K_x, f \rangle$ where $\bar{p}$ is the costate. Applying the maximum principle, we obtain the differential equation $\dot{\bar{p}}^* = -\overline{H}_x\big|_* = -(f_x)^T\big|_* \bar{p}^* - p_0^* K_{xx}\big|_* f\big|_* - p_0^*(f_x)^T\big|_* K_x\big|_*$ with the boundary condition $\bar{p}^*(t_f) = 0$. The latter tells us, by the way, that $p_0^* \neq 0$, hence from now on we assume p_0^* and $\bar{p}^*$ to have been normalized so that $p_0^* = -1$. Now, matching the expression for $\overline{H}$ with (4.44), let us define the costate for our original problem to be

$$p^*(t) := \bar{p}^*(t) - K_x(x^*(t)). \tag{4.45}$$

Its final value is consistent with (4.43). We can check that p^* defined in this way satisfies the correct canonical equation.

Exercise 4.7   *Verify that p^* given by (4.45) satisfies the differential equation $\dot{p}^* = -H_x\big|_*$ with respect to the Hamiltonian (4.44). Show also that $p^*(t) \neq 0$ for all t.* $\square$

By construction, H is maximized and equals 0 along the optimal trajectory because these properties hold for $\overline{H}$. We thus conclude that the statement of the maximum principle for the present case, with respect to the Hamiltonian (4.44), is obtained from that for the Basic Variable-Endpoint Control Problem by eliminating p_0^* and changing the transversality condition to $p^*(t_f) = -K_x(x^*(t_f))$. Recall that we already saw such a boundary condition for the costate in (3.36).

Exercise 4.8   *Consider the following more general problem: the system is $\dot{x} = f(t, x, u)$, the cost is $J(u) = \int_{t_0}^{t_f} L(t, x(t), u(t))dt + K(x_f)$, and the target set is $S = [t_0, \infty) \times S_1$ where S_1 is a k-dimensional surface in $\mathbb{R}^n$ (the terminal time t_f is free). Reduce this problem by a change of variables to the Basic Variable-Endpoint Control Problem and arrive at a precise statement of the maximum principle for it.*   □

INITIAL SETS

We now want to mention how the maximum principle can be extended in another direction. We usually assume that the initial state x_0 is fixed, while the final state x_f may vary within some set S_1. Here, let us briefly consider the possibility that x_0 may vary as well, so that we have an initial *set* instead of a fixed initial state. We can impose separate constraints on x_0 and x_f or, more generally, we can require that $\begin{pmatrix} x_0 \\ x_f \end{pmatrix}$ belong to some surface S_2 in $\mathbb{R}^{2n}$. The latter formulation allows us to capture situations where there is a joint constraint on x_0 and x_f; for example, if the trajectory is to be closed—i.e., the initial state is free but the trajectory must return to it—then S_2 is the "diagonal" in $\mathbb{R}^{2n}$. The terminal time t_f can be either free or fixed, as before.

It turns out that this more general set-up can be easily handled by modifying the transversality condition, which will now involve the values of the costate both at the initial time and at the final time. Namely, the transversality condition will now say that the vector $\begin{pmatrix} p^*(t_0) \\ -p^*(t_f) \end{pmatrix}$ must be orthogonal to the tangent space to S_2 at $\begin{pmatrix} x^*(t_0) \\ x^*(t_f) \end{pmatrix}$:

$$\left\langle \begin{pmatrix} p^*(t_0) \\ -p^*(t_f) \end{pmatrix}, d \right\rangle = 0 \qquad \forall d \in T_{\begin{pmatrix} x^*(t_0) \\ x^*(t_f) \end{pmatrix}} S_2. \qquad (4.46)$$

The total number of boundary conditions for the system of canonical equations is still $2n$, since each additional degree of freedom for $x^*(t_0)$ leads to one additional constraint on $p^*(t_0)$.

Exercise 4.9   *Give a heuristic argument in support of the transversality condition (4.46), based on an appropriate modification of the proof of the maximum principle.*   □

4.4   TIME-OPTIMAL CONTROL PROBLEMS

Time-optimal control problems not only provide a useful illustration of the maximum principle, but also have several important features that make them interesting in their own right. Among these are the bang-bang principle and connections with Lie brackets, to be discussed in this section. We organize the material so as to progress from more specific to more general settings, and we begin by revisiting Example 1.1 from Section 1.1.

4.4.1   Example: double integrator

Consider the system

$$\ddot{x} = u, \qquad u \in [-1, 1] \tag{4.47}$$

which can represent a car with position $x \in \mathbb{R}$ and with bounded acceleration u acting as the control (negative acceleration corresponds to braking). Let us study the problem of "parking" the car at the origin, i.e., bringing it to rest at $x = 0$, in minimal time. It is clear—and will follow from the analysis we give below—that the system can indeed be brought to rest at the origin from every initial condition. However, since the control is bounded, we cannot do this arbitrarily fast (we are ignoring the trivial case when the system is initialized at the origin). Thus we expect that there exists an optimal control u^* which achieves the transfer in the smallest amount of time. (As we have already mentioned, the issue of existence of optimal controls is important and nontrivial in general; it will be treated in Section 4.5.)

We know that the dynamics of the double integrator (4.47) are equivalently described by the state-space equations

$$\begin{aligned} \dot{x}_1 &= x_2, \\ \dot{x}_2 &= u \end{aligned} \tag{4.48}$$

where we assume that the initial values $x_1(t_0)$ and $x_2(t_0)$ are given. The running cost is $L \equiv 1$ (cf. Example 3.2 in Section 3.2 where we discussed another time-optimal control problem). Accordingly, the Hamiltonian is $H = p_1 x_2 + p_2 u + p_0$. Let u^* be an optimal control. Our problem is a special case of the Basic Fixed-Endpoint Control Problem, and we now apply the maximum principle to characterize u^*. The costate $p^* = \begin{pmatrix} p_1^* \\ p_2^* \end{pmatrix}$ must satisfy the adjoint equation

$$\begin{pmatrix} \dot{p}_1^* \\ \dot{p}_2^* \end{pmatrix} = \begin{pmatrix} -H_{x_1}|_* \\ -H_{x_2}|_* \end{pmatrix} = \begin{pmatrix} 0 \\ -p_1^* \end{pmatrix}. \tag{4.49}$$

The first line of (4.49) implies that p_1^* is equal to a constant, say c_1. The second line of (4.49) then says that p_2^* is given by $p_2^*(t) = -c_1 t + c_2$, where c_2 is another constant.

Next, from the Hamiltonian maximization condition and the fact that $U = [-1, 1]$ we have

$$u^*(t) = \operatorname{sgn}(p_2^*(t)) := \begin{cases} 1 & \text{if } p_2^*(t) > 0, \\ -1 & \text{if } p_2^*(t) < 0, \\ ? & \text{if } p_2^*(t) = 0. \end{cases} \qquad (4.50)$$

The third case in (4.50) is meant to indicate that when p_2^* equals 0, making the Hamiltonian independent of u, the value of u^* can in principle be arbitrary. (By our convention that u^* is continuous either from the left or from the right everywhere, we know that u^* actually cannot take any values other than 1 or -1.) Can p_2^* be identically 0 over some time interval? If this happens, then $\dot{p}_2^* = -p_1^*$ must also be identically 0 on that interval. However, we already saw (on page 130) that p^* cannot vanish in a time-optimal control problem, because the fact that $H \equiv 0$ (statement 3 of the maximum principle) would then imply that $p_0^* = 0$ and the nontriviality condition would be violated. Therefore, p_2^* may only equal 0 at isolated points in time, and the formula (4.50) defines u^* uniquely everywhere away from these times. How many zero crossings can p_2^* have? We derived a little while ago that p_2^* is a linear function of time. Thus p_2^* can cross the value 0 *at most once*.

We conclude that the optimal control u^* takes only the values ± 1 and switches between these values at most once. Interpreted in terms of bringing a car to rest at the origin, the optimal control strategy consists in switching between maximal acceleration and maximal braking. The initial sign and the switching time of course depend on the initial condition. The property that u^* only switches between the extreme values ± 1 is intuitively natural and important; such controls are called *bang-bang*.

It turns out that for the present problem, the pattern identified above uniquely determines the optimal control law for every initial condition. To see this, let us plot the solutions of the system (4.47) in the $(x, \dot{x})$-plane for $u \equiv \pm 1$. For $u \equiv 1$, repeated integration gives $\dot{x}(t) = t + a$ and then $x(t) = \frac{1}{2}t^2 + at + b$ for some constants a and b. The resulting relation $x = \frac{1}{2}\dot{x}^2 + c$ (where $c = b - a^2/2$) defines a family of parabolas in the $(x, \dot{x})$-plane parameterized by $c \in \mathbb{R}$. Similarly, for $u \equiv -1$ we obtain the family of parabolas $x = -\frac{1}{2}\dot{x}^2 + c$, $c \in \mathbb{R}$. These curves are shown in Figure 4.15 (a,b), with the arrows indicating the direction in which they are traversed. It is easy to see that only two of these trajectories hit the origin (which is the prescribed final point). Their union is the thick curve in Figure 4.15 (c), which we call the *switching curve* and denote by Γ; it is defined by the relation $x = -\frac{1}{2}|\dot{x}|\dot{x}$. The optimal control strategy thus consists in applying $u = 1$ or $u = -1$ depending on whether the initial point is below or above Γ, then switching the control value exactly on Γ and subsequently following Γ to the origin; no switching is needed if the initial point is already on Γ. (Thinking slightly differently, we can generate all possible optimal trajectories—which cover the entire plane—by starting at the origin and flowing backward in time, first following Γ and then switching at an arbitrary point on Γ.) Recalling the interpretation of our problem as that of parking a car using bounded acceleration/braking, the reader can easily relate the optimal trajectories in the $(x, \dot{x})$-plane with the corresponding motions of the car along the x-axis. Note that if the car is initially moving away from the origin, then it begins braking until it stops, turns around, and starts accelerating (this is a "false" switch because u actually remains constant), and then u switches sign and the car starts braking again.

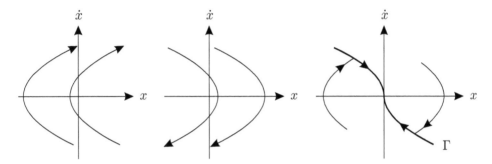

Figure 4.15: Bang-bang time-optimal control of the double integrator: (a) trajectories for $u \equiv 1$, (b) trajectories for $u \equiv -1$, (c) the switching curve and optimal trajectories

The optimal control law that we just found has two important features. First, as we already said, it is bang-bang. Second, we see that it can be described in the form of a *state feedback law*. This is interesting because in general, the maximum principle only provides an open-loop description of an optimal control; indeed, $u^*(t)$ depends, besides the state $x^*(t)$, on the costate $p^*(t)$, but we managed to eliminate this latter dependence here. It is natural to ask for what more general classes of systems time-optimal controls have these two properties, i.e., are bang-bang and take the state feedback form. The bang-bang property will be examined in detail in the next two subsections. The problem of representing optimal controls as state feedback laws is rather intricate and will not be treated in this book, except for the two exercises below.

Exercise 4.10 *Suppose that we modify the above problem by removing the requirement that the final velocity be 0. In other words, instead of bringing the car to rest at the origin, we now want to simply hit the origin with an arbitrary speed ("slam into a wall"). Use the maximum principle to find the optimal control. Is it bang-bang? Can you express it as a state feedback and arrive at a complete description of the optimal trajectories in the $(x, \dot{x})$-plane? Include an explanation of the car's optimal motions.* $\square$

The next exercise is along the same lines but the solution is less obvious.

Exercise 4.11 *Consider the problem of bringing to rest at the origin in minimal time the state of the forced harmonic oscillator $\ddot{x} + x = u$ using controls satisfying $u \in [-1, 1]$. Answer the same questions as in Exercise 4.10.* □

4.4.2 Bang-bang principle for linear systems

Consider now a system with general linear time-invariant dynamics

$$\dot{x} = Ax + Bu \tag{4.51}$$

where $x \in \mathbb{R}^n$ and $u \in U \subset \mathbb{R}^m$. We want to investigate under what conditions time-optimal controls for this system have a bang-bang property along the lines of what we saw in the previous example. Of course, we need to specify what the control set U is and exactly what we mean by a bang-bang property for controls taking values in this U. As a natural generalization of the interval $[-1, 1] \subset \mathbb{R}$ to higher dimensions, we take U to be an m-dimensional hypercube:

$$U = \{u \in \mathbb{R}^m : u_i \in [-1, 1], \, i = 1, \ldots, m\}. \tag{4.52}$$

This is a reasonable control set representing independent control actuators. We take the magnitude constraints on the different components of u to be the same just for simplicity; the extension to different constraints is immediate.

Suppose that the control objective is to steer x from a given initial state x_0 to a given final state x_1 in minimal time. To be sure that this problem is well posed, we assume that there exists some control u that achieves the transfer from x_0 to x_1 (in some time). As we will see in Section 4.5 (Theorem 4.3), this guarantees that a time-optimal control $u^* : [t_0, t^*] \to U$ exists. We now use the maximum principle to characterize it. The Hamiltonian is $H(x, u, p, p_0) = \langle p, Ax + Bu \rangle + p_0$. The Hamiltonian maximization condition implies that

$$\langle p^*(t), Bu^*(t) \rangle = \max_{u \in U} \langle p^*(t), Bu(t) \rangle \tag{4.53}$$

for all $t \in [t_0, t^*]$. We can rewrite this formula in terms of the input components as

$$\sum_{i=1}^{m} \langle p^*(t), b_i \rangle u_i^*(t) = \max_{u \in U} \sum_{i=1}^{m} \langle p^*(t), b_i \rangle u_i$$

where $b_1, \ldots, b_m$ are the columns of B. Since the components $u_i^*(t)$ of the optimal control can be chosen independently, it is clear that each term in the summation must be maximized:

$$\langle p^*(t), b_i \rangle u_i^*(t) = \max_{|u_i| \leq 1} \langle p^*(t), b_i \rangle u_i, \qquad i = 1, \ldots, m.$$

It has now become obvious that we must have

$$u_i^*(t) = \operatorname{sgn}(\langle p^*(t), b_i \rangle) = \begin{cases} 1 & \text{if } \langle p^*(t), b_i \rangle > 0, \\ -1 & \text{if } \langle p^*(t), b_i \rangle < 0, \\ ? & \text{if } \langle p^*(t), b_i \rangle = 0 \end{cases} \qquad (4.54)$$

for each i. Observe that (4.54) resembles (4.50). Similarly to how we proceeded from that formula, we now need to investigate the possibility that $\langle p^*, b_i \rangle \equiv 0$ on some interval of time for some i, as this would prevent us from determining u_i^* on that interval.

The adjoint equation is $\dot{p}^* = -A^T p^*$, which gives $p^*(t) = e^{A^T(t^*-t)} p^*(t^*)$. From this we obtain $\langle p^*(t), b_i \rangle = \langle p^*(t^*), e^{A(t^*-t)} b_i \rangle$. This is a real analytic function of t; hence, if it vanishes on some time interval, then it vanishes for all t, together with all its derivatives. Calculating these derivatives at $t = t^*$, we arrive at the equalities

$$\langle p^*(t^*), b_i \rangle = \langle p^*(t^*), A b_i \rangle = \cdots = \langle p^*(t^*), A^{n-1} b_i \rangle = 0. \qquad (4.55)$$

As we know, $p^*(t^*)$ cannot be 0 in a time-optimal control problem (see page 130). Since (4.55) means orthogonality of $p^*(t^*)$ to the vectors b_i, $A b_i$, ..., $A^{n-1} b_i$, we can rule it out by assuming that these vectors span $\mathbb{R}^n$, i.e., that (A, b_i) is a controllable pair. We need this to be true for each i; in other words, the system should be controllable with respect to each individual input channel. Such linear control systems are called *normal*.

Let us now collect the properties of the optimal control u^* that we are able to derive under the above normality assumption. None of the functions $\langle p^*(\cdot), b_i \rangle$ equal 0 on any time interval; being real analytic functions, they only have finitely many zeros on the interval $[t_0, t^*]$. Using the formula (4.54), we see that each function u_i^* only takes the values ± 1 and switches between these values finitely many times. Away from these switching times, u_i^* is uniquely determined by (4.54). We conclude that the overall optimal control u^* takes values only in the set of *vertices* of the hypercube U, has finitely many discontinuities (switches), and is unique everywhere else. Generalizing the earlier notion, we say that controls taking values in the set of vertices of U are *bang-bang* (or have the bang-bang property); the result that we have just obtained is a version of the **bang-bang principle** for linear systems.

Before closing this discussion, it is instructive to see how the above bang-bang property can be established in a self-contained way, without relying on the maximum principle. More precisely, the argument outlined next essentially rederives the maximum principle from scratch for the particular problem at hand (in the spirit of Section 4.3.1). Solutions of the system (4.51) take the form

$$x(t) = e^{A(t-t_0)} x_0 + \int_{t_0}^{t} e^{A(t-s)} B u(s) ds.$$

For $t \geq t_0$, let us introduce the set of points reachable from $x(t_0) = x_0$ at time t:

$$R^t(x_0) := \left\{ e^{A(t-t_0)}x_0 + \int_{t_0}^t e^{A(t-s)}Bu(s)ds : u(s) \in U, \, t_0 \leq s \leq t \right\}. \quad (4.56)$$

We know that

$$x_1 = e^{A(t^*-t_0)}x_0 + \int_{t_0}^{t^*} e^{A(t^*-s)}Bu^*(s)ds \in R^{t^*}(x_0). \quad (4.57)$$

In fact, the optimal time t^* is the smallest time t such that $x_1 \in R^t(x_0)$. It follows that x_1 must belong to the *boundary* of the reachable set $R^{t^*}(x_0)$; indeed, if it were an interior point of $R^{t^*}(x_0)$ then we could reach it sooner. We will see a little later (in Section 4.5) that the set $R^{t^*}(x_0)$ is compact and convex. Along the lines of Section 4.2.7, there exists a hyperplane that passes through x_1 and contains $R^{t^*}(x_0)$ on one side; such a hyperplane is said to *support* $R^{t^*}(x_0)$ at x_1. Denoting a suitably chosen normal vector to this hyperplane by $p^*(t^*)$, we have $\langle p^*(t^*), x_1 \rangle \geq \langle p^*(t^*), x \rangle$ for all $x \in R^{t^*}(x_0)$. Using (4.56) and (4.57), we easily obtain that

$$\int_{t_0}^{t^*} \langle p^*(t^*), e^{A(t^*-s)}Bu^*(s) \rangle ds \geq \int_{t_0}^{t^*} \langle p^*(t^*), e^{A(t^*-s)}Bu(s) \rangle ds$$

for all $u : [t_0, t^*] \to U$, which in view of the formula $e^{A^T(t^*-s)}p^*(t^*) = p^*(s)$ is equivalent to

$$\int_{t_0}^{t^*} \langle p^*(s), Bu^*(s) \rangle ds \geq \int_{t_0}^{t^*} \langle p^*(s), Bu(s) \rangle ds \qquad \forall \, u : [t_0, t^*] \to U.$$

From this it is not difficult to recover the fact that (4.53) must hold for (almost) all $t \in [t_0, t^*]$, and we can proceed from there as before.

Exercise 4.12 *Consider the same time-optimal control problem as above, but take U to be the unit ball in $\mathbb{R}^m$ (with respect to the standard Euclidean norm). Suppose that (A, B) is a controllable pair (do not assume normality). What can you say about optimal controls?* □

The assumption of normality, which was needed to prove the bang-bang property of time-optimal controls for U a hypercube, is quite strong. A different, weaker version of the bang-bang principle could be formulated as follows. Rather than wishing for *every* time-optimal control to be bang-bang, we could ask whether every state x_1 reachable from x_0 by some control is also reachable from x_0 *in the same time* by a bang-bang control; in other words, whether reachable sets for bang-bang controls coincide with reachable sets for all controls. This would imply that, even though not all time-optimal controls are necessarily bang-bang, we can always select one that is bang-bang. It turns out that this modified bang-bang principle holds for every linear control system (no controllability assumption is necessary) and every control set U that is a convex polyhedron. The proof requires a refinement of the above argument and some additional steps; see [Sus83, Section 8.1] for details.

4.4.3 Nonlinear systems, singular controls, and Lie brackets

Let us now investigate whether the preceding results can be extended beyond the class of linear control systems. Regarding the bang-bang principle cited in the previous paragraph, the hope that it might be true for general nonlinear systems is quickly shattered by the following example.

Example 4.1 *For the planar system*

$$\dot{x}_1 = x_2^2 - 1,$$
$$\dot{x}_2 = u \qquad\qquad (4.58)$$

with the control constraint $u \in [-1, 1]$, consider the problem of going from $\begin{pmatrix} 1 \\ 0 \end{pmatrix}$ to $\begin{pmatrix} 0 \\ 0 \end{pmatrix}$ in minimal time. The reader should be able to solve it by inspection. The unique optimal control is $u^* \equiv 0$. Indeed, it accomplishes the desired transfer along the x_1-axis in time 1, while any other control would make x_2 deviate from 0, slowing down the decrease of x_1 and resulting in a transfer time larger than 1. Note that 0 is an interior point of the control set $U = [-1, 1]$, hence u^* is not a bang-bang control.

Let us verify that this conclusion is in agreement with the maximum principle. The Hamiltonian is $H = p_1(x_2^2 - 1) + p_2 u + p_0$. By the Hamiltonian maximization condition, the optimal control must satisfy the same relation (4.50) as in the example of Section 4.4.1. We see that for u^* to take the value 0 (or any other value different from ± 1), p_2^* must vanish. Turning our attention to the adjoint equation, we have $\dot{p}_1^* = 0$ and $\dot{p}_2^* = -2p_1^* x_2^*$. Compared with Section 4.4.1, this situation is a bit more complicated because the p^*-dynamics and x^*-dynamics are coupled. We know, however, that the optimal trajectory is completely contained in the x_1-axis, i.e., $x_2^* \equiv 0$ along the optimal trajectory. Thus we have $\dot{p}_2^* \equiv 0$ and so p_2^* is a constant. The value of this constant is determined by the terminal condition $p^*(1)$. Since we are dealing with a fixed-endpoint problem, there are no constraints on $p^*(1)$. In particular, $p_2^*(1)$ can be 0 in which case $p_2^* \equiv 0$. Note that p_1^* can still be nonzero, and there is no contradiction with the maximum principle. □

A distinguishing feature of the above example is that the function p_2^*, whose sign determines the value of the optimal control u^*, identically vanishes. Consequently, the Hamiltonian maximization condition alone does not give us enough information to find u^*. In problems where this situation occurs on some interval of time, the optimal control on that interval is called *singular*, and the corresponding piece of the optimal state trajectory is called a *singular arc*.

Example 4.1 should not be taken to suggest, however, that we must give up the hope of formulating a bang-bang principle for nonlinear systems. After all, we saw in Section **4.4.2** that even for linear systems, to be able to prove that all time-optimal controls are bang-bang we need the normality assumption. It is conceivable that the bang-bang property of time-optimal controls for certain nonlinear systems can be guaranteed under an appropriate nonlinear counterpart of that assumption.

Motivated by these remarks, our goal now is to better formalize the phenomenon of singularity—and reach a deeper understanding of its reasons—for a class of systems that includes the linear systems considered in Section **4.4.2** as well as the nonlinear system (4.58). This class is composed of nonlinear systems affine in controls, defined as

$$\dot{x} = f(x) + G(x)u = f(x) + \sum_{i=1}^{m} g_i(x)u_i \qquad (4.59)$$

where $x \in \mathbb{R}^n$, $u \in U \subset \mathbb{R}^m$, $G(x)$ is an $n \times m$ matrix whose columns are $g_1(x), \ldots, g_m(x)$, and for the control set U we again take the hypercube (4.52). The Hamiltonian for the time-optimal control problem is

$$H(x, u, p, p_0) = \left\langle p, f(x) + \sum_{i=1}^{m} g_i(x)u_i \right\rangle + p_0.$$

From the Hamiltonian maximization condition we obtain, completely analogously to Section **4.4.2**, that the components $u_i^*(t)$ of the optimal control are determined by the signs of the functions $\varphi_i(t) := \langle p^*(t), g_i(x^*(t)) \rangle$. These functions of time (always associated with a specific optimal trajectory) are called the *switching functions*. To investigate the bang-bang property, we need to study zeros of the switching functions.

In order to simplify calculations, from this point on we assume that $m = 1$, so that the input u is scalar and we have only one switching function

$$\varphi(t) = \langle p^*(t), g(x^*(t)) \rangle. \qquad (4.60)$$

The optimal control satisfies

$$u^*(t) = \operatorname{sgn}(\varphi(t)) = \begin{cases} 1 & \text{if } \varphi(t) > 0, \\ -1 & \text{if } \varphi(t) < 0, \\ ? & \text{if } \varphi(t) = 0. \end{cases} \qquad (4.61)$$

The canonical equations are $\dot{x}^* = f(x^*) + g(x^*)u^*$ and

$$\dot{p}^* = -H_x|_* = -(f_x)^T|_* p^* - (g_x)^T|_* p^* u^*$$

where f_x and g_x are the Jacobian matrices of f and g. Let us now compute the derivative of φ:

$$\dot{\varphi} = \langle \dot{p}^*, g(x^*) \rangle + \langle p^*, g_x(x^*)\dot{x}^* \rangle = -\langle (f_x)^T p^*, g \rangle|_* - \langle (g_x)^T p^*, g \rangle|_* u^*$$
$$+ \langle p^*, g_x f \rangle|_* + \langle p^*, g_x g \rangle|_* u^* = \langle p^*, g_x f - f_x g \rangle|_*. \tag{4.62}$$

We see that $\dot{\varphi}$ is the inner product of p^* with the vector $(g_x f - f_x g)|_*$. Perhaps the vector field $g_x f - f_x g$, which we have not encountered up to now, has some significant meaning?

Example 4.2 *Let us take f and g to be linear vector fields: $f(x) = Ax$ and $g(x) = Bx$ for some $n \times n$ matrices A and B. In this case (4.59) reduces to a* bilinear *(not linear!) control system $\dot{x} = Ax + Bxu$. We have $f_x \equiv A$, $g_x \equiv B$, and $(g_x f - f_x g)(x) = BAx - ABx = [B, A]x$, where $[B, A] := BA - AB$ is the* commutator, *or Lie bracket, of B and A.* □

In general, the *Lie bracket* of two differentiable vector fields f and g is another vector field defined as

$$[f, g](x) := g_x(x)f(x) - f_x(x)g(x).$$

Note that the definitions of the Lie bracket for matrices (in linear algebra) and for vector fields (in differential geometry) usually follow the opposite sign conventions. The geometric meaning of the Lie bracket—which justifies its alternative name "commutator"—is as follows (see Figure 4.16). Suppose that, starting at some point x_0, we move along the vector field f for ε units of time, then along the vector field g for ε units of time, after that along $-f$ (backward along f) for ε units of time, and finally along $-g$ for ε units of time. It is straightforward (although quite tedious) to check that for small ε the resulting motion is approximated, up to terms of higher order in ε, by $\varepsilon^2 [f, g](x_0)$. In particular, we will return to x_0 if $[f, g] \equiv 0$ in a neighborhood of x_0, in which case we say that f and g *commute*.

We can now write the result of the calculation (4.62) more informatively as

$$\dot{\varphi}(t) = \langle p^*(t), [f, g](x^*(t)) \rangle. \tag{4.63}$$

Coupled with the law (4.61), this equation reveals a fundamental connection between Lie brackets and optimal control.

Exercise 4.13 *For the multiple-input system (4.59), compute $\dot{\varphi}_i$, $i = 1, \ldots, m$. Express the result in terms of Lie brackets.* □

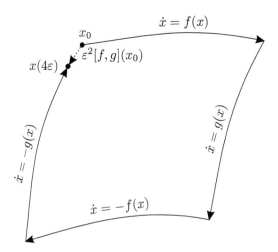

Figure 4.16: Geometric interpretation of the Lie bracket

Lie brackets can help us shed light on the bang-bang property. For a singular optimal control to exist, φ must identically vanish on some time interval. In view of (4.60) and (4.63), this can happen only if $p^*(t)$ stays orthogonal to both $g(x^*(t))$ and $[f,g](x^*(t))$. We have seen that in time-optimal problems $p^*(t) \neq 0$ for all t. Thus for planar systems ($n = 2$) we can rule out singularity if g and $[f,g]$ are linearly independent along the optimal trajectory.

If $n > 2$ or g and $[f,g]$ are not linearly independent, then we have to look at higher derivatives of φ and see what it takes for them to vanish as well. Rather than differentiating $\dot{\varphi}$ again, let us revisit our derivation of $\dot{\varphi}$ and try to see a general pattern in it. Consider an arbitrary differentiable vector field h on $\mathbb{R}^n$. Following the same calculation steps as in (4.62) and using the definition of the Lie bracket, we easily arrive at

$$\frac{d}{dt}\langle p^*(t), h(x^*(t))\rangle = \langle p^*(t), [f,h](x^*(t))\rangle + \langle p^*(t), [g,h](x^*(t))\rangle u^*(t). \quad (4.64)$$

The formula (4.63) for $\dot{\varphi}$ is recovered from this result as a special case by setting $h := g$ which gives $[g,h] = [g,g] = 0$. Now, if we want to compute $\ddot{\varphi}$, we only need to set $h := [f,g]$ to obtain the following expression in terms of iterated Lie brackets of f and g:

$$\ddot{\varphi}(t) = \langle p^*(t), [f,[f,g]](x^*(t))\rangle + \langle p^*(t), [g,[f,g]](x^*(t))\rangle u^*(t). \quad (4.65)$$

A singular optimal control must make $\ddot{\varphi}$ vanish. The control

$$u^*(t) = -\frac{\langle p^*(t), [f,[f,g]](x^*(t))\rangle}{\langle p^*(t), [g,[f,g]](x^*(t))\rangle} \quad (4.66)$$

can potentially be singular if $\langle p^*, g(x^*) \rangle = \langle p^*, [f, g](x^*) \rangle \equiv 0$. However, it should meet the magnitude constraint $|u^*(t)| \leq 1$. If we assume, for example, that the relation

$$[g, [f, g]](x) = \alpha(x)g(x) + \beta(x)[f, g](x) + \gamma(x)[f, [f, g]](x)$$

holds with $|\gamma(x)| < 1$ for all x, then (4.66) would not be an admissible control unless $\langle p^*, [f, [f, g]](x^*) \rangle \equiv 0$. To investigate the possibility that this last function does vanish, we need to consider its derivative given by (4.64) with $h := [f, [f, g]]$, and so on.

Exercise 4.14 *Continuing this process, generate a set of conditions that rule out the existence of singular optimal controls. You might find it convenient to write these conditions in terms of the operators* $(\operatorname{ad} f)^k$, $k = 0, 1, \ldots$ *defined by* $(\operatorname{ad} f)^0(g) := g$ *and* $(\operatorname{ad} f)^k(g) := [f, (\operatorname{ad} f)^{k-1}(g)]$ *for* $k \geq 1$. $\square$

We are now in a position to gain a better insight into our earlier observations by using the language of Lie brackets.

LINEAR SYSTEMS (SECTION 4.4.2). In the single-input case, we have $f(x) = Ax$ and $g(x) = b$. Calculating the relevant Lie brackets, we obtain $[f, g] = -Ab$, $[f, [f, g]] = A^2 b$, $[g, [f, g]] = 0$, $[f, [f, [f, g]]] = -A^3 b$, $[g, [f, [f, g]]] = 0$, etc. A crucial consequence of linearity is that iterated Lie brackets containing two g's are 0, which makes the derivatives of the switching function φ independent of u. It is easy to see that φ cannot vanish if the vectors $b, Ab, \ldots, A^{n-1}b$ span $\mathbb{R}^n$, which is precisely the controllability condition.

EXAMPLE 4.1 REVISITED. For the system (4.58), we have $f = \begin{pmatrix} x_2^2 - 1 \\ 0 \end{pmatrix}$ and $g = \begin{pmatrix} 0 \\ 1 \end{pmatrix}$. The first Lie bracket is

$$[f, g](x) = - \begin{pmatrix} 0 & 2x_2 \\ 0 & 0 \end{pmatrix} \begin{pmatrix} 0 \\ 1 \end{pmatrix} = \begin{pmatrix} -2x_2 \\ 0 \end{pmatrix}.$$

On the x_1-axis, where the singular optimal trajectory lives, $[f, g]$ vanishes and so g and $[f, g]$ do not span $\mathbb{R}^2$. In fact, $\langle p^*, g(x^*) \rangle = \langle p^*, [f, g](x^*) \rangle = 0$ when $p_2^* = 0$. The next Lie bracket that we should then calculate is

$$[f, [f, g]](x) = \begin{pmatrix} 0 & -2 \\ 0 & 0 \end{pmatrix} \begin{pmatrix} x_2^2 - 1 \\ 0 \end{pmatrix} - \begin{pmatrix} 0 & 2x_2 \\ 0 & 0 \end{pmatrix} \begin{pmatrix} -2x_2 \\ 0 \end{pmatrix} = \begin{pmatrix} 0 \\ 0 \end{pmatrix}.$$

Since $u \equiv 0$, (4.65) gives $\ddot{\varphi} \equiv 0$. All higher-order Lie brackets are obviously 0, hence $\varphi \equiv 0$. We see that all information about the singularity is indeed encoded in the Lie brackets.

It is worth noting that singular controls are not necessarily complicated. The optimal control $u^* \equiv 0$ in Example 4.1 is actually quite simple. For single-input planar systems $\dot{x} = f(x) + g(x)u$, $x \in \mathbb{R}^2$, $u \in [-1, 1]$, with f and g real analytic, it can be shown that all time-optimal trajectories are concatenations of a *finite* number of "bang" pieces (each corresponding to either $u = 1$ or $u = -1$) and real analytic singular arcs. It is natural to ask whether a similar claim holds for other optimal control problems in $\mathbb{R}^2$ or for time-optimal control problems in $\mathbb{R}^3$. We are about to see that these two questions are related and that the answer to both is negative.

4.4.4   Fuller's problem

Consider again the double integrator (4.48) with the same control constraint $u \in [-1, 1]$. Let the cost functional be

$$J(u) = \int_{t_0}^{t_f} x_1^2(t) dt \tag{4.67}$$

and let the target set be $S = [t_0, \infty) \times \left\{ \begin{pmatrix} 0 \\ 0 \end{pmatrix} \right\}$. In other words, the final time is free and the final state is the origin, but this is not a minimal-time problem any more. The Hamiltonian is $H = p_1 x_2 + p_2 u + p_0 x_1^2$ and the optimal control must again satisfy (4.50), i.e., the switching function is $\varphi(t) = p_2^*(t)$. The adjoint equation is

$$\begin{aligned} \dot{p}_1^* &= -2p_0^* x_1^*, \\ \dot{p}_2^* &= -p_1^*. \end{aligned} \tag{4.68}$$

We claim that singular arcs are ruled out. Indeed, along a singular arc p_2^* must vanish. Inspecting the differential equations for p_2^*, p_1^*, and x_1^*, we see that p_1^*, x_1^*, and x_2^* must then vanish too. Thus the only possible singular arc is the trivial one consisting of the equilibrium at the origin.

   It follows that all optimal controls are bang-bang, with switches occurring when p_2^* equals 0. Our findings up to this point replicate those in the time-optimal setting of Section **4.4.1**. There, we used the fact that p_2^* depended linearly on t to go further and show that the optimal control has at most one switch. The present adjoint equation (4.68) is different from (4.49) and so we can no longer reach the same conclusion. Instead, it turns out that the optimal solution has the following properties (which we state without proof):

1) Optimal controls are bang-bang with *infinitely many switches*.

2) Switching takes place on the curve $\left\{ \begin{pmatrix} x_1 \\ x_2 \end{pmatrix} : x_1 + \gamma |x_2| x_2 = 0 \right\}$ where $\gamma \approx 0.445$.

3) Time intervals between consecutive switches decrease in geometric progression.

The last property guarantees that the final time is finite. The occurrence of a switching pattern in which switching times form an infinite sequence accumulating near the final time is known as *Fuller's phenomenon* (or *Zeno behavior*, or "chattering"). Note that while bang-bang controls with finitely many switches are piecewise continuous (in fact, piecewise constant), optimal controls for Fuller's problem are only measurable (see page 86). Figure 4.17 shows the switching curve[6] and a sketch of an optimal state trajectory.

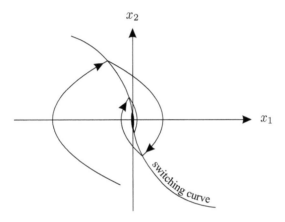

Figure 4.17: An optimal trajectory for Fuller's problem

We note that the switching curves in Fuller's problem and in the time-optimal control problem for the double integrator (treated in Section 4.4.1) are given by the same formula, but in the time-optimal problem we had a different value of γ, namely, $\gamma = 1/2$. The nature of switching, however, is drastically different in the two cases. We can embed both problems in the parameterized family of problems with the cost functional $J(u) = \int_{t_0}^{t_f} |x_1(t)|^\nu dt$ where $\nu \geq 0$ is a parameter. For $\nu = 0$ we recover the time-optimal problem while for $\nu = 2$ we recover Fuller's problem. Interestingly, one can prove that there exists a "bifurcation value" $\bar\nu \approx 0.35$ with the following property: for $\nu \in [0, \bar\nu]$ the optimal control is bang-bang with at most one switch, while for $\nu > \bar\nu$ we obtain Fuller's phenomenon.

The next exercise shows that Fuller's phenomenon is also observed in time-optimal control problems starting with dimension 3. Its solution can be based on Fuller's problem.

Exercise 4.15 *Give an example of a control system $\dot x = f(x) + g(x)u$ with $x \in \mathbb{R}^3$ and $u \in [-1, 1]$ for which transferring the state to the origin in*

---

[6]The shape of the switching curve in the figure is slightly distorted for better visualization; in reality the trajectory converges to 0 faster.

minimal time involves infinitely many switches between $u = 1$ and $u = -1$ (at least for some initial states x_0).   □

4.5   EXISTENCE OF OPTIMAL CONTROLS

To motivate the subject of this section, let us consider the following (obviously absurd) claim: *The largest positive integer is 1.* To "prove" this statement, let N be the largest positive integer and suppose that $N \neq 1$. Then we have $N^2 > N$, which contradicts the property of N being the largest positive integer. Therefore, $N = 1$.

This argument is known as *Perron's paradox.* Although clearly farcical, it highlights a serious issue which we have been dodging up until now: it warns us about the danger of assuming the existence of an optimal solution. Indeed, finding the largest positive integer is an optimization problem. Of course, a solution to this problem does not exist. In the language of this book, the above reasoning correctly shows that $N = 1$ is a *necessary condition* for optimality. Thus a necessary condition can be useless—even misleading—unless we know that a solution exists.

We know very well that the maximum principle only provides necessary conditions for optimality. The same is true for the Euler-Lagrange equation and several other results that we have derived along the way. We have said repeatedly that fulfillment of necessary conditions alone does not guarantee optimality. However, the basic question of whether an optimal solution even exists has not been systematically addressed yet, and it is time to do it now.

Suppose that we are given an optimal control problem which falls into the general class of problems formulated in Section 3.3. How can we ensure the existence of an optimal control? For a start, we must assume that there exists at least one control u that drives the state of the system from the initial condition to the target set S (more precisely, the pair $(t, x(t))$ should evolve from (t_0, x_0) to S). Otherwise, the problem is ill posed (has infinite cost). We can view this as a type of controllability assumption. It may be nontrivial to check, especially for general control sets U. Still, it would be nice if this controllability assumption were enough to guarantee the existence of an optimal control. Simple examples show that unfortunately this is not the case.

Example 4.3   *For the standard integrator $\dot{x} = u$, with $x, u \in \mathbb{R}$, consider the problem of steering x from $x_0 = 0$ to $x_1 = 1$ in minimal time. An optimal solution does not exist: it is easy to see that arbitrarily small positive transfer times are achievable, but accomplishing the transfer in time 0 is impossible.*   □

Observe that in the above example, arbitrarily fast transfer is possible because the control set $U = \mathbb{R}$ is unbounded. Let us see if the problem becomes well posed when U is taken to be bounded.

Example 4.4 *Consider the same problem as in Example 4.3 except let $U = [0, 1)$. Now the state can be transferred from 0 to 1 in an arbitrary time larger than 1, but we cannot achieve the transfer in time 1. Hence, an optimal solution again does not exist.* □

This time, the difficulty is caused by the fact that the control set is not closed. The reader is probably beginning to guess that we want to have both boundedness and closedness, i.e., *compactness*. To make this statement precise, however, we should be working not with the control set U itself but with the set of all points reachable from x_0 using controls that take values in U. For $t \geq t_0$, let us denote by $R^t(x_0)$ the set of points reachable from $x(t_0) = x_0$ at time t (for a given control set U). We already worked with reachable sets on page 140, in the context of linear systems. We saw there that if $t^* - t_0$ is the fastest transfer time from x_0 to x_1, then x_1 must be a boundary point of $R^{t^*}(x_0)$; this situation is illustrated in Figure 4.18.

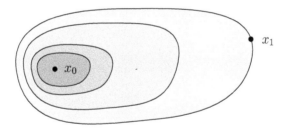

Figure 4.18: Propagation of reachable sets $R^t(x_0)$, $t_0 \leq t \leq t^*$

In the two scalar examples considered earlier, the reachable set computation is straightforward and the negative conclusions are readily explained by the above property being impossible to satisfy. In Example 4.3, $R^t(x_0) = \mathbb{R}$ for all $t > 0$, which is unbounded and cannot contain $x_1 = 1$ on its boundary. In Example 4.4, $R^1(x_0) = [0, 1)$ which is not closed and does not include the boundary point 1, while for each $t > 1$ the set $R^t(x_0)$ already contains 1 in its interior. We can now formulate a refined guess saying that for optimal controls to exist, we want to have compact reachable sets. We could generalize this discussion to target sets instead of fixed terminal points; we will also see that compactness of reachable sets is relevant not only for time-optimal control problems.

For nonlinear systems, explicit computation of reachable sets is usually not feasible. Instead, we can rely on the following general result known as **Filippov's theorem**: *Given a control system in the standard form (3.17) with $u \in U$, assume that its solutions exist on a time interval $[t_0, t_f]$ for*

*all controls $u(\cdot)$ and that for every pair (t, x) the set $\{f(t, x, u) : u \in U\}$
is compact and convex. Then the reachable set $R^t(x_0)$ is compact for each
$t \in [t_0, t_f]$.* Filippov's theorem actually has the following stronger form,
which implies compactness of $R^t(x_0)$: Under the above assumptions, the set
of all trajectories of the system, as a subset of $\mathcal{C}^0([t_0, t_f], \mathbb{R}^n)$, is compact
with respect to the topology induced by the 0-norm $\|\cdot\|_0$ (i.e., the topology
of uniform convergence).

We do not prove Filippov's theorem but make some comments on it. For
the result to be valid, controls must a priori be understood as measurable
functions from $[t_0, t_f]$ to U; see the discussion on page 86. (We can hope to
show afterwards that *optimal* controls belong to a nicer class of functions,
e.g., that they enjoy the bang-bang property.) It is important to note that
the first hypothesis in the theorem (existence of solutions on a given interval)
does not follow from the second hypothesis (compactness and convexity of
the right-hand side); for example, solutions of $\dot{x} = x^2 + u$ blow up in finite
time even for $U = \{0\}$. Boundedness of the reachable sets can be shown
without the convexity assumption, but convexity is crucial for establishing
closedness (the argument relies on the separating hyperplane theorem). The
next exercise should help clarify the role of the convexity assumption.

Exercise 4.16   *Consider the planar system*

$$\dot{x}_1 = u,$$
$$\dot{x}_2 = x_1^2$$

*with $x_0 = \begin{pmatrix} 0 \\ 0 \end{pmatrix}$ and $U = \{-1, 1\}$. Show that the reachable sets are not closed.
Use this fact to define a cost functional such that the resulting optimal control
problem has no solution. Explain how the situation changes if we "convexify"
the problem by redefining U to be the interval $[-1, 1]$.*   □

Filippov's theorem applies to useful classes of systems, most notably, to
systems affine in controls given by (4.59) with compact and convex control
sets U. For such systems, the set $\{f(x) + G(x)u : u \in U\}$ is compact and
convex for each x as the image of U under an affine map. As we already
warned the reader, existence of solutions on a time interval of interest needs
to be assumed separately. In the special case of the linear system (4.51),
though, global existence of solutions is automatic; the right-hand side of
the system is allowed to depend on time as well. For linear systems the
reachable set $R^t(x_0)$ is also convex, as can be easily seen from convexity of
U and the variation-of-constants formula for solutions.

Filippov's theorem provides a sufficient condition for compactness of
reachable sets. Earlier, we argued that compactness of reachable sets should
be useful for proving existence of optimal controls. Let us now confirm that

this is indeed true, at least for certain classes of problems. The connection between compactness of reachable sets and existence of optimal controls is especially easy to see for problems in the Mayer form (terminal cost only). Suppose that the control system satisfies the hypotheses of Filippov's theorem, that the cost is $J(u) = K(x_f)$ with K a continuous function, and—in order to arrive at the simplest case—that the target set is $S = \{t_1\} \times \mathbb{R}^n$ (a fixed-time, free-endpoint problem). Since K is a continuous function and $R^{t_1}(x_0)$ is a compact set, the Weierstrass Theorem (see page 9) guarantees that K has a global minimum over $R^{t_1}(x_0)$. By the definition of $R^{t_1}(x_0)$, there exists at least one control that steers the state to this minimum at time t_1, and every such control is optimal.

For certain fixed-time problems in the Bolza form, it is possible to establish existence of optimal controls by combining compactness of the set of system trajectories (provided by the stronger form of Filippov's theorem) with continuity of the cost functional on this set and invoking the infinite-dimensional version of the Weierstrass Theorem (see page 23). Alternatively, one can reduce the problem to the Mayer form (as in Section 3.3.2) and then use the previous result. In general, the argument needs to be tailored to a particular type of problem at hand. We will now show in detail how existence of optimal controls can be proved for linear time-optimal control problems. The next result says that compactness and convexity of U and a controllability assumption (the existence of some control achieving the desired transfer) is all we need for an optimal control to exist. (As we know from Section 4.4.2, such an optimal control is automatically bang-bang if U is a hypercube, or can be chosen to be bang-bang if U is an arbitrary convex polyhedron.)

THEOREM 4.3 (Existence of time-optimal controls for linear systems). *Consider the linear control system* (4.51) *with a compact and convex control set U. Let the control objective be to steer x from a given initial state $x(t_0) = x_0$ to a given final state x_1 in minimal time. Assume that $x_1 \in R^t(x_0)$ for some $t \geq t_0$. Then there exists a time-optimal control.*

PROOF. Let $t^* := \inf\{t \geq t_0 : x_1 \in R^t(x_0)\}$. The time t^* is well defined because by the theorem's hypothesis the set over which the infimum is being taken is nonempty. (We note that, in view of boundedness of U, we have $t^* > t_0$ unless $x_1 = x_0$.) We will be done if we show that $x_1 \in R^{t^*}(x_0)$, i.e., that t^* is actually a minimum.[7] This will mean that there exists a control u^* that transfers x_0 to x_1 at time t^*, and by the definition of t^* no control does it faster.

By the definition of infimum, there exists a sequence $t_k \searrow t^*$ such that $x_1 \in R^{t_k}(x_0)$ for each k. Then, by the definition of $R^{t_k}(x_0)$, for each k there

---

[7]This situation is to be contrasted with the one in Example 4.4.

is a control u_k such that

$$x_1 = e^{A(t_k - t_0)}x_0 + \int_{t_0}^{t_k} e^{A(t_k - s)}Bu_k(s)ds.$$

The above equation shows that the same point x_1 belongs to the reachable sets $R^{t_k}(x_0)$ for different k. To be able to use the closedness property of reachable sets guaranteed by Filippov's theorem, we would rather work with different points belonging to the same reachable set. To this end, let us truncate the trajectories corresponding to the controls u_k at time t^* and define, for each k,

$$x^k := e^{A(t^* - t_0)}x_0 + \int_{t_0}^{t^*} e^{A(t^* - s)}Bu_k(s)ds. \qquad (4.69)$$

All these points by construction belong to the same reachable set $R^{t^*}(x_0)$.

We now claim that the sequence of points x^k converges to x_1. If we establish this fact then the proof will be finished, because the closedness of the reachable set $R^{t^*}(x_0)$ will imply that $x_1 \in R^{t^*}(x_0)$ as needed. To prove the convergence claim, let us write

$$x_1 - x^k = \left(e^{A(t_k - t^*)} - I\right)e^{A(t^* - t_0)}x_0 + \left(e^{A(t_k - t^*)} - I\right)\int_{t_0}^{t^*} e^{A(t^* - s)}Bu_k(s)ds$$

$$+ \int_{t^*}^{t_k} e^{A(t_k - s)}Bu_k(s)ds = \left(e^{A\varepsilon_k} - I\right)x^k + \int_0^{\varepsilon_k} e^{A(\varepsilon_k - \tau)}Bu_k(t^* + \tau)d\tau$$

where we defined $\varepsilon_k := t_k - t^*$ and made the substitution $s = t^* + \tau$. Taking the norm on both sides, we have

$$|x_1 - x^k| \le \left\|e^{A\varepsilon_k} - I\right\||x^k| + \varepsilon_k \max\left\{|e^{A\delta}Bu| : \delta \in [0, \varepsilon_k], u \in U\right\} \quad (4.70)$$

where the maximum is well defined because U is compact. Similarly, we can take the norm on both sides of (4.69) to obtain the bound

$$|x^k| \le \left\|e^{A(t^* - t_0)}\right\||x_0| + (t^* - t_0)\max\left\{|e^{A\delta}Bu| : \delta \in [0, t^* - t_0], u \in U\right\}$$

which is independent of k, implying that the sequence $|x^k|$ is uniformly bounded. Using this fact in (4.70) and noting that $\varepsilon_k \to 0$ hence also $\|e^{A\varepsilon_k} - I\| \to 0$, we conclude that $|x_1 - x^k|$ converges to 0 as claimed. $\qquad \square$

Existence results are also available for other, more general classes of control problems (see Section 4.6 for further information). Although the proofs of such results are usually not constructive, the knowledge that a solution exists rules out the risks suggested by Perron's paradox and provides a basis for applying the maximum principle—which, as we have seen, often allows one to actually find an optimal control.

4.6 NOTES AND REFERENCES FOR CHAPTER 4

The Basic Fixed-Endpoint Control Problem and Basic Variable-Endpoint Control Problem match the Special Problem 1 and Special Problem 2 in [AF66], respectively, and the statements of the maximum principle for these two problems correspond to Theorem 5-5P and Theorem 5-6P in [AF66, Section 5-15]. Our proof of the maximum principle is also heavily based on the one presented in [AF66], but there are significant differences between the two proofs. First, we substantially reorganized the proof structure of [AF66], changing the main steps and the order in which they are given. Second, we filled in some details of proving Lemmas 4.1 and 4.2 not included in [AF66]; these are taken from the proofs of Lemmas 3 and 10 in the original book [PBGM62]. Finally, as we already explained earlier, our sign convention (maximum versus minimum) is the opposite of that in [AF66]. Among other expositions of the proof of the maximum principle built around similar ideas, we note the one in [LM67] and the more modern approach of [Sus00]. Our derivation of the variational equation (step 4) proceeded similarly to the argument in [Kha02, Section 3.3].

A precise definition of the topology mentioned in Section 4.3 (uniform convergence for x and $\mathcal{L}_1$ convergence for u), which leads to a notion of local optimality suitable for the maximum principle, can be found in [MO98, p. 23]; that book uses the terminology "convergence in Pontryagin's sense" and "Pontryagin minimum." The book [Vin00] works with the somewhat different concept of a $W^{1,1}$ local minimizer (defined there on pp. 287–288). Changes of variables for deriving the maximum principle for other classes of problems, including those discussed in Section 4.3.1, are thoroughly covered in [AF66]. For the reader wishing to reconstruct a proof of the maximum principle for the Mayer problem in more detail, Section 6.1 of [BP07] should be of help. In the majority of the literature the initial state is assumed to be fixed; references that do deal with variable initial states and resulting transversality conditions include [LM67] and [Vin00].

The example of Section 4.4.1 is standard and appears in many places, including [PBGM62], [BP07, Section 7.3], [AF66, Section 7-2], [Kno81], and [Son98, Chapter 10]. The last three references also discuss the bang-bang principle for linear systems derived in Section 4.4.2. When all eigenvalues of the matrix A are real, a more precise result can be obtained, namely, each component of the optimal control can switch at most $n-1$ times (see, e.g., [AF66, Theorem 6-8]). For further information on synthesizing optimal (in particular, time-optimal) controls in state feedback form, see [Sus83, PS00, BP04] and the references therein. The weaker bang-bang principle mentioned at the end of Section 4.4.2 remains true for linear time-varying systems and for arbitrary compact and convex control sets U, provided that by bang-bang controls one understands controls taking values in the set of extreme points of U; see, e.g., [LM67, Ces83, Kno81] (the last of which only addresses the time-varying aspect). However, the proofs become less elementary. Another advantage of U being a convex polyhedron is that a bound on the number of switches (which in general depends on the length of the time interval) can be established, as explained in [Sus83, Section 8.1]. Our treatment of singular optimal controls and their connection with Lie brackets in Section 4.4.3 was inspired by the nice expositions in [Sus83] and [Sus00, Handout 5]. The characterization of time-optimal controls in the plane mentioned at the end of Section 4.4.3 is stated in [Sus83] as Theorem 8.4.1. Fuller's problem is studied in [Ful85] and in the earlier work by the same author cited in that paper. The relevant results are conveniently summarized in [Rya87], while a detailed analysis can be found in [Jur96, Chapter 10]. Fuller's phenomenon is observed in other problems too, such as controlling a Dubins car (see [AS04, Section 20.6]). The paper [Neu03] describes how it can play a role in design of marine reserves for optimal harvesting.[8]

Perron's paradox and its ramifications are carefully examined in [You80]. Examples 4.3 and 4.4 are contained in [Son98, Remark 10.1.2]. Filippov's theorem is commonly attributed to [Fil88, §7, Theorem 3], although that book cites earlier works. An extra step (the so-called Filippov's Selection Lemma) is needed to pass from the differential inclusion setting of [Fil88] to that of control systems; see, e.g., [Vin00, Section 2.3]. Filippov's theorem is also discussed, among many other sources, in [BP07, Section 3.5]. Boundedness of reachable sets without the convexity assumption is shown in [LSW96, Proposition 5.1]. For linear systems, compactness of reachable sets is addressed in a more direct fashion in [Son98, Section 10.1]. Existence of optimal controls for the Mayer problem with more general target sets, as well as for the Bolza problem via a reduction to the Mayer problem, is investigated in [BP07, Chapter 5]. Our proof of Theorem 4.3 follows [Son98] and [Kno81]. This argument can be extended to nonlinear systems affine in controls; the essential details are in [Son98, Lemma 10.1.2 and Remark 10.1.10] and [Sus79, pp. 633–634]. The assumption of convexity of U in Theorem 4.3 can actually be dropped, as a consequence of available bang-bang theorems for linear systems (see [BP07, Theorem 5.1.2] or [Sus83, Theorem 8.1.1]). Further results on existence of optimal controls can be found in [Kno81] and [LM67]. For an in-depth treatment of this issue we recommend the book [Ces83] which dedicates several chapters to it. The issue of controllability of linear systems with bounded controls, which is relevant to the material of Sections 4.4.2 and 4.5, is considered in [Jur96, Chapter 5] (a necessary condition for controllability is that the matrix A has purely imaginary eigenvalues).

We must stress that the basic setting of this chapter and the maxi-

---

[8]We thank Patrick De Leenheer for acquainting us with this interesting application.

mum principle that we developed are far from being the most general possible. As we said earlier, we work with piecewise continuous controls rather than the larger class of measurable controls considered in [LM67], [Vin00], and [Sus00]. An analogous maximum principle is valid for Lipschitz and not necessarily $\mathcal{C}^1$ systems, but its derivation requires additional technical tools; see [Cla89], [Vin00], [Sus07], and the references therein. Although the classical formulation of the maximum principle relies on first-order analysis, high-order versions are also available; see [Kre77], [Bre85], and [AS04, Chapter 20]. The applicability of the maximum principle has been extended to other problems; some of these—namely, control problems on manifolds and hybrid control problems—will be touched upon in Chapter 7, while others are completely beyond the scope of this book (for example, see [YZ99] for a stochastic maximum principle). Versions of the maximum principle for discrete-time systems exist and are surveyed in [NRV84]; see also [Bol78].

Chapter Five

The Hamilton-Jacobi-Bellman Equation

5.1 DYNAMIC PROGRAMMING AND THE HJB EQUATION

Right around the time when the maximum principle was being developed in the Soviet Union, on the other side of the Atlantic Ocean (and of the iron curtain) Bellman wrote the following in his book [Bel57]: "In place of determining the optimal sequence of decisions from the *fixed* state of the system, we wish to determine the optimal decision to be made at *any* state of the system. Only if we know the latter, do we understand the intrinsic structure of the solution." The approach realizing this idea, known as *dynamic programming*, leads to necessary as well as sufficient conditions for optimality expressed in terms of the so-called Hamilton-Jacobi-Bellman (HJB) partial differential equation for the optimal cost. These concepts are the subject of the present chapter. Developed independently from—even, to some degree, in competition with—the maximum principle during the cold war era, the resulting theory is very different from the one presented in Chapter 4. Nevertheless, both theories have their roots in calculus of variations and there are important connections between the two, as we will explain in Section 5.2 (see also Section 7.2).

5.1.1 Motivation: the discrete problem

The dynamic programming approach is quite general, but to fix ideas we first present it for the purely discrete case. Consider a system of the form

$$x_{k+1} = f(x_k, u_k), \qquad k = 0, 1, \ldots, T - 1$$

where x_k lives in a finite set X consisting of N elements, u_k lives in a finite set U consisting of M elements, and T, N, M are fixed positive integers. We suppose that each possible transition from some x_k to x_{k+1} corresponding to some control value u_k has a cost assigned to it, and there is also a terminal cost function on X. For each trajectory, the total cost accumulated at time T is the sum of the transition costs at time steps $0, \ldots, T - 1$ plus the terminal cost at x_T. For a given initial state x_0 we want to minimize this total cost, the terminal state x_T being free.

The most naive approach to this problem is as follows: starting from x_0, enumerate all possible trajectories going forward up to time T, calculate the cost for each one, then compare them and select the optimal one. Figure 5.1 provides a visualization of this scenario. It is easy to estimate the computational effort required to implement such a solution: there are M^T possible trajectories and we need T additions to compute the cost for each one, which results in roughly $O(M^T T)$ algebraic operations.

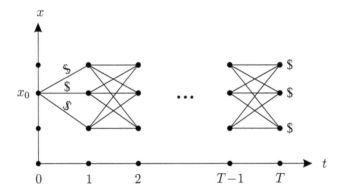

Figure 5.1: Discrete case: going forward

We now examine an alternative approach, which might initially appear counterintuitive: let us go backward in time. At $k = T$, terminal costs are known for each x_k. At $k = T - 1$, for each x_k we find to which x_{k+1} we should jump so as to have the smallest cost (the one-step running cost plus the terminal cost). Write this optimal "cost-to-go" next to each x_k and mark the selected path (see Figure 5.2). In case of more than one path giving the same cost, choose one of them at random. Repeat these steps for $k = T - 2, \ldots, 0$, working with the costs-to-go computed previously in place of the terminal costs.

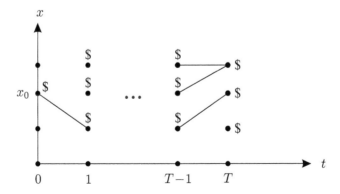

Figure 5.2: Discrete case: going backward

We claim that when we are done, we will have generated an optimal trajectory from each x_0 to some x_T. The justification of this claim relies on the principle of optimality, an observation that we already made during the proof of the maximum principle (see page 108). In the present context this principle says that for each time step k, if x_k is a point on an optimal trajectory then the remaining decisions (from time k onward) must constitute an optimal policy with respect to x_k as the initial condition. What the principle of optimality does for us here is guarantee that the paths we discard going backward cannot be portions of optimal trajectories. On the other hand, in the previous approach (going forward) we are not able to discard any paths until we reach the terminal time and finish the calculations.

Let us assess the computational effort associated with this backward scheme. At each time k, for each state x_k and each control u_k we need to add the cost of the corresponding transition to the cost-to-go already computed for the resulting x_{k+1}. Thus, the number of required operations is $O(NMT)$. Comparing this with the $O(M^T T)$ operations needed for the earlier forward scheme, we conclude that the backward computation is more efficient for large T, with N and M fixed. Of course, the number of operations will still be large if N and M are large (this is the "curse of dimensionality").

Actually, the above comparison is not really accurate because the backward scheme provides much more information: it finds the optimal policy for every initial condition x_0, and in fact it tells us what the optimal decision is at every x_k for all k. We can restate this last property as follows: the backward scheme yields the optimal control policy in the form of a *state feedback law*. In the forward scheme, on the other hand, to handle all initial conditions we would need $O(NM^T T)$ operations, and we would still not cover all states x_k for $k > 0$; hence, a state feedback is not obtained. We see that the backward scheme fulfills the objective formulated in Bellman's quote at the beginning of this chapter. This recursive scheme serves as an example of the general method of dynamic programming.

5.1.2   Principle of optimality

We now return to the continuous-time optimal control problem that we have been studying since Section 3.3, defined by the control system (3.17) and the Bolza cost functional (3.20). For concreteness, assume that we are dealing with a fixed-time, free-endpoint problem, i.e., the target set is $S = \{t_1\} \times \mathbb{R}^n$. (Handling other target sets requires some modifications, on which we briefly comment in what follows.) We can then write the cost functional as

$$J(u) = \int_{t_0}^{t_1} L(t, x(t), u(t)) dt + K(x(t_1)).$$

As we already remarked in Section 3.3.2, a more accurate notation for this functional would be $J(t_0, x_0, u)$ as it depends on the initial data.

The basic idea of dynamic programming is to consider, instead of the problem of minimizing $J(t_0, x_0, u)$ for given t_0 and x_0, the *family* of minimization problems associated with the cost functionals

$$J(t, x, u) = \int_t^{t_1} L(s, x(s), u(s))ds + K(x(t_1)) \qquad (5.1)$$

where t ranges over $[t_0, t_1]$ and x ranges over $\mathbb{R}^n$; here $x(\cdot)$ on the right-hand side denotes the state trajectory corresponding to the control $u(\cdot)$ and satisfying $x(t) = x$. (There is a slight abuse of notation here; the second argument x of J in (5.1) is a fixed point, and only the third argument u is a function of time.) In accordance with Bellman's roadmap, our goal is to derive a *dynamic relationship* among these problems, and ultimately to solve *all of them*.

To this end, let us introduce the *value function*

$$V(t, x) := \inf_{u_{[t, t_1]}} J(t, x, u) \qquad (5.2)$$

where the notation $u_{[t, t_1]}$ indicates that the control u is restricted to the interval $[t, t_1]$. Loosely speaking, we can think of $V(t, x)$ as the optimal cost (cost-to-go) from (t, x). It is important to note, however, that the existence of an optimal control—and hence of the optimal cost—is not actually assumed, which is why we work with an infimum rather than a minimum in (5.2). If an optimal control exists, then the infimum turns into a minimum and V coincides with the optimal cost-to-go. In general, the infimum need not be achieved, and might even equal $-\infty$ for some (t, x).

It is clear that the value function must satisfy the boundary condition

$$V(t_1, x) = K(x) \qquad \forall\, x \in \mathbb{R}^n. \qquad (5.3)$$

In particular, if there is no terminal cost ($K \equiv 0$) then we have $V(t_1, x) = 0$. The boundary condition (5.3) is of course a consequence of our specific problem formulation. If the problem involved a more general target set $S \subset [t_0, \infty) \times \mathbb{R}^n$, then the boundary condition would read $V(t, x) = K(x)$ for $(t, x) \in S$.

The basic principle of dynamic programming for the present case is a continuous-time counterpart of the principle of optimality formulated in Section 5.1.1, already familiar to us from Chapter 4. Here we can state this property as follows, calling it again the **principle of optimality**: *For every* $(t, x) \in [t_0, t_1) \times \mathbb{R}^n$ *and every* $\Delta t \in (0, t_1 - t]$, *the value function* V *defined in (5.2) satisfies the relation*

$$V(t, x) = \inf_{u_{[t, t + \Delta t]}} \left\{ \int_t^{t + \Delta t} L(s, x(s), u(s))ds + V(t + \Delta t, x(t + \Delta t)) \right\} \quad (5.4)$$

where $x(\cdot)$ on the right-hand side is the state trajectory corresponding to the control $u_{[t,t+\Delta t]}$ and satisfying $x(t) = x$. The intuition behind this statement is that to search for an optimal control, we can search over a small time interval for a control that minimizes the cost over this interval plus the subsequent optimal cost-to-go. Thus the minimization problem on the interval $[t, t_1]$ is split into two, one on $[t, t + \Delta t]$ and the other on $[t + \Delta t, t_1]$; see Figure 5.3.

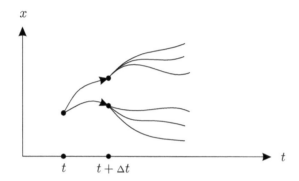

Figure 5.3: Continuous time: principle of optimality

The above principle of optimality may seem obvious. However, it is important to justify it rigorously, especially since we are using an infimum and not assuming existence of optimal controls. We give "one half" of the proof by verifying that

$$V(t, x) \geq \overline{V}(t, x) \tag{5.5}$$

where $\overline{V}(t, x)$ denotes the right-hand side of (5.4):

$$\overline{V}(t, x) := \inf_{u_{[t,t+\Delta t]}} \left\{ \int_t^{t+\Delta t} L(s, x(s), u(s)) ds + V(t + \Delta t, x(t + \Delta t)) \right\}.$$

By (5.2) and the definition of infimum, for every $\varepsilon > 0$ there exists a control u_ε on $[t, t_1]$ such that

$$V(t, x) + \varepsilon \geq J(t, x, u_\varepsilon). \tag{5.6}$$

Writing x_ε for the corresponding state trajectory, we have

$$J(t, x, u_\varepsilon) = \int_t^{t+\Delta t} L(s, x_\varepsilon(s), u_\varepsilon(s)) ds + J(t + \Delta t, x_\varepsilon(t + \Delta t), u_\varepsilon)$$

$$\geq \int_t^{t+\Delta t} L(s, x_\varepsilon(s), u_\varepsilon(s)) ds + V(t + \Delta t, x_\varepsilon(t + \Delta t)) \geq \overline{V}(t, x)$$

where the two inequalities follow directly from the definitions of V and $\overline{V}$, respectively. Since (5.6) holds with an arbitrary $\varepsilon > 0$, the desired inequality (5.5) is established.

Exercise 5.1 *Complete the proof of the principle of optimality by showing the reverse inequality $V(t, x) \leq \overline{V}(t, x)$.*   □

5.1.3   HJB equation

In the principle of optimality (5.4) the value function V appears on both sides with different arguments. We can thus think of (5.4) as describing a dynamic relationship among the optimal values of the costs (5.1) for different t and x, which we declared earlier to be our goal. However, this relationship is rather clumsy and not very convenient to use in its present form. What we will now do is pass to its more compact infinitesimal version, which will take the form of a partial differential equation (PDE). The steps that follow rely on first-order Taylor expansions; the reader will recall that we used somewhat similar calculations when deriving the maximum principle. First, write $x(t + \Delta t)$ appearing on the right-hand side of (5.4) as

$$x(t + \Delta t) = x + f(t, x, u(t))\Delta t + o(\Delta t)$$

where we remembered that $x(t) = x$. This allows us to express $V(t+\Delta t, x(t+\Delta t))$ as

$$V(t + \Delta t, x(t + \Delta t)) = V(t, x) + V_t(t, x)\Delta t + \langle V_x(t, x), f(t, x, u(t))\Delta t \rangle + o(\Delta t)$$
$$(5.7)$$

(for now we proceed under the assumption—whose validity we will examine later—that V is $\mathcal{C}^1$). We also have

$$\int_t^{t+\Delta t} L(s, x(s), u(s))ds = L(t, x, u(t))\Delta t + o(\Delta t). \qquad (5.8)$$

Substituting the expressions given by (5.7) and (5.8) into the right-hand side of (5.4), we obtain

$$V(t, x) = \inf_{u_{[t,t+\Delta t]}} \big\{ L(t, x, u(t))\Delta t + V(t, x)$$
$$+ V_t(t, x)\Delta t + \langle V_x(t, x), f(t, x, u(t))\Delta t \rangle + o(\Delta t) \big\}.$$

The two $V(t, x)$ terms cancel out (because the one inside the infimum does not depend on u and can be pulled outside), which leaves us with

$$0 = \inf_{u_{[t,t+\Delta t]}} \big\{ L(t, x, u(t))\Delta t + V_t(t, x)\Delta t + \langle V_x(t, x), f(t, x, u(t))\Delta t \rangle + o(\Delta t) \big\}.$$
$$(5.9)$$

Let us now divide by Δt and take it to be small. In the limit as $\Delta t \to 0$ the higher-order term $o(\Delta t)/\Delta t$ disappears, and the infimum is taken over the instantaneous value of u at time t (in fact, already in (5.9) the control values $u(s)$ for $s > t$ affect the expression inside the infimum only through the $o(\Delta t)$ term). Pulling $V_t(t,x)$ outside the infimum as it does not depend on u, we conclude that the equation

$$\boxed{-V_t(t,x) = \inf_{u \in U} \left\{ L(t,x,u) + \langle V_x(t,x), f(t,x,u) \rangle \right\}} \qquad (5.10)$$

must hold for all $t \in [t_0, t_1)$ and all $x \in \mathbb{R}^n$. This equation for the value function is called the **Hamilton-Jacobi-Bellman (HJB) equation**. It is a PDE since it contains partial derivatives of V with respect to t and x. The accompanying boundary condition is (5.3).

Note that the terminal cost appears only in the boundary condition and not in the HJB equation itself. In fact, the specifics on the terminal cost and terminal time did not play a role in our derivation of the HJB equation. For different target sets, the boundary condition changes (as we already discussed) but the HJB equation remains the same. However, the HJB equation will not hold for $(t,x) \in S$ just like it does not hold at $t = t_1$ in the fixed-time case, because the principle of optimality is not valid there.

We can apply one more transformation in order to rewrite the HJB equation in a simpler—and also more insightful—way. It is easy to check that (5.10) is equivalent to

$$V_t(t,x) = \sup_{u \in U} \{ \langle -V_x(t,x), f(t,x,u) \rangle - L(t,x,u) \}. \qquad (5.11)$$

Let us now recall our earlier definition (3.28) of the Hamiltonian, reproduced here:

$$H(t,x,u,p) := \langle p, f(t,x,u) \rangle - L(t,x,u).$$

We see that the expression inside the supremum in (5.11) is nothing but the Hamiltonian, with $-V_x$ playing the role of the costate. This brings us to the Hamiltonian form of the HJB equation:

$$V_t(t,x) = \sup_{u \in U} H\left(t,x,u,-V_x(t,x)\right). \qquad (5.12)$$

So far, the existence of an optimal control has not been assumed. When an optimal (in the global sense) control u^* does exist, the infimum in the previous calculations can be replaced by a minimum and this minimum is achieved when u^* is plugged in. In particular, the principle of optimality (5.4) yields

$$V(t, x^*(t)) = \min_{u_{[t,t+\Delta t]}} \left\{ \int_t^{t+\Delta t} L(s, x(s), u(s)) ds + V(t + \Delta t, x(t + \Delta t)) \right\}$$

$$= \int_t^{t+\Delta t} L(s, x^*(s), u^*(s)) ds + V(t + \Delta t, x^*(t + \Delta t))$$

where $x^*(\cdot)$ and $x(\cdot)$ are trajectories corresponding to $u^*(\cdot)$ and $u(\cdot)$, respectively, both passing through the same point $x^*(t)$ at time t. From this, repeating the same steps that led us earlier to the HJB equation (5.10), we obtain

$$-V_t(t, x^*(t)) = \min_{u \in U} \{L(t, x^*(t), u) + \langle V_x(t, x^*(t)), f(t, x^*(t), u) \rangle\}$$

$$= L(t, x^*(t), u^*(t)) + \langle V_x(t, x^*(t)), f(t, x^*(t), u^*(t)) \rangle . \tag{5.13}$$

Expressed in terms of the Hamiltonian, the second equation in (5.13) becomes

$$H(t, x^*(t), u^*(t), -V_x(t, x^*(t))) = \max_{u \in U} H(t, x^*(t), u, -V_x(t, x^*(t))). \quad (5.14)$$

This Hamiltonian maximization condition is analogous to the one we had in the maximum principle. We see that if we can find a closed-form expression for the control that maximizes the Hamiltonian—or, equivalently, the control that achieves the infimum in the HJB equation (5.10)—then the HJB equation becomes simpler and more explicit.

Example 5.1   *Consider the standard integrator $\dot{x} = u$ (with $x, u \in \mathbb{R}$) and let $L(x, u) = x^4 + u^4$. The corresponding HJB equation is*

$$-V_t(t, x) = \inf_{u \in \mathbb{R}} \{x^4 + u^4 + V_x(t, x)u\}. \tag{5.15}$$

Since the expression inside the infimum is polynomial in u, we can easily find the control that achieves the infimum: differentiating with respect to u, we have $4u^3 + V_x(t, x) = 0$ which yields $u = -\left(\frac{1}{4}V_x(t, x)\right)^{1/3}$. Plugging this control into the HJB equation (5.15), we obtain

$$-V_t(t, x) = x^4 - 3\left(\tfrac{1}{4}V_x(t, x)\right)^{4/3}. \tag{5.16}$$

Assuming that we can solve the PDE (5.16) for the value function V, the optimal control is given in the state feedback form[1] $u^*(t) = -\left(\frac{1}{4}V_x(t, x^*(t))\right)^{1/3}$. Solving (5.16), however, appears to be difficult.   □

Example 5.2   *Consider again the minimal-time parking problem from Section 4.4.1. The HJB equation for this problem is*

$$-V_t(t, x) = \inf_{u \in [-1, 1]} \{1 + V_{x_1}(t, x)x_2 + V_{x_2}(t, x)u\}. \tag{5.17}$$

---

[1]Note the difference between the previous expression for the minimizing control u, which is given pointwise in t and x, and the current expression for u^* which is evaluated along the corresponding state trajectory x^*. The claim about optimality of u^* is actually somewhat informal at this point, but will be carefully justified very soon (in Section 5.1.4).

Note that this is a free-time, fixed-endpoint problem, for which the boundary condition takes the form $V(t, 0) = 0$ for all t, and (5.17) is valid away from $x = 0$. The infimum on the right-hand side of (5.17) is achieved by setting

$$u = -\text{sgn}(V_{x_2}(t, x)) = \begin{cases} 1 & \text{if } V_{x_2}(t, x) < 0, \\ -1 & \text{if } V_{x_2}(t, x) > 0, \\ ? & \text{if } V_{x_2}(t, x) = 0. \end{cases}$$

It is informative to compare this expression for u with (4.50). Substituting it into (5.17), we arrive at the simplified HJB equation

$$-V_t(t, x) = 1 + V_{x_1}(t, x)x_2 - |V_{x_2}(t, x)|. \qquad (5.18)$$

The optimal control is given by the feedback law $u^*(t) = -\text{sgn}(V_{x_2}(t, x^*(t)))$, whose implementation of course hinges on our ability to solve (5.18) for V. We end the example here but invite the reader to play more with it in the next exercise. □

Exercise 5.2 *Do you see a way to further simplify the HJB equation (5.18) from Example 5.2? Can you solve this HJB equation and find the value function? Can you use the above analysis to reproduce the results we obtained in Section 4.4.1 (i.e., the bang-bang principle and the complete description of the optimal feedback law)? Conversely, can you use those earlier results to obtain more information about the value function?* □

INFINITE-HORIZON PROBLEM

The reader may find it somewhat discouraging that even for simple examples like the ones we just presented, solving the HJB equation analytically is a challenging task. Nevertheless, the HJB equation gives a considerable insight into the problem, and one can often apply numerical techniques to obtain approximate solutions. We now discuss one important situation in which the HJB equation takes a simpler form. Suppose that both the control system and the cost functional are time-invariant, i.e., $f = f(x, u)$ and $L = L(x, u)$, and that there is no terminal cost ($K \equiv 0$). Keeping the final state free, we let the final time t_1 approach ∞; in the limit, the cost functional becomes $J(u) = \int_{t_0}^{\infty} L(x(t), u(t))dt$ and we have what is called an *infinite-horizon* problem.[2] It is clear that in this scenario, the cost does not depend on the

---

[2]For this infinite-horizon problem to be well posed, we need to be sure that the cost is finite at least for some controls. We will investigate this issue in detail for the linear quadratic regulator problem in the next chapter. For now we proceed formally, without worrying about the possibility that the problem might be ill posed.

initial time, hence the value function depends on x only: $V = V(x)$. Thus the partial derivative V_t vanishes and the HJB equation (5.10) reduces to

$$0 = \inf_{u \in U} \left\{ L(x, u) + \langle V_x(x), f(x, u) \rangle \right\}. \tag{5.19}$$

We let the reader think of other problem formulations for which the HJB equation simplifies in the same way.

The PDE (5.19) may still be difficult to solve, but it is certainly more tractable than the general HJB equation (5.10). In the special case when x is scalar, (5.19) is actually an ODE. Let us consider the infinite-horizon version of Example 5.1. The HJB equation becomes $x^4 - 3\left(\frac{1}{4}V_x(x)\right)^{4/3} = 0$ from which we derive $V_x(x) = \pm\left(\frac{1}{3}\right)^{3/4} 4x^3$. We must choose the plus sign because V should be positive definite (since L is positive definite). We do not even need to solve for V because the optimal feedback law is obtained from V_x directly: $u^*(t) = -\left(\frac{1}{4}V_x(x^*(t))\right)^{1/3} = -\left(\frac{1}{3}\right)^{1/4} x^*(t)$.

5.1.4 Sufficient condition for optimality

Together, the HJB equation—written as (5.10) or (5.12)—and the Hamiltonian maximization condition (5.14) constitute necessary conditions for optimality. It should be clear that all we proved so far is their necessity. Indeed, defining V to be the value function, we showed that it must satisfy the HJB equation. Assuming further that an optimal control exists, we showed that it must maximize the Hamiltonian along the optimal trajectory. However, we will see next that these conditions are also sufficient for optimality. Namely, we will establish the following **sufficient condition for optimality**: *Suppose that a C^1 function $\widehat{V} : [t_0, t_1] \times \mathbb{R}^n \to \mathbb{R}$ satisfies the HJB equation*

$$-\widehat{V}_t(t, x) = \inf_{u \in U} \left\{ L(t, x, u) + \langle \widehat{V}_x(t, x), f(t, x, u) \rangle \right\} \tag{5.20}$$

(for all $t \in [t_0, t_1)$ and all $x \in \mathbb{R}^n$) and the boundary condition

$$\widehat{V}(t_1, x) = K(x). \tag{5.21}$$

Suppose that a control $\hat{u} : [t_0, t_1] \to U$ and the corresponding trajectory $\hat{x} : [t_0, t_1] \to \mathbb{R}^n$, with the given initial condition $\hat{x}(t_0) = x_0$, satisfy everywhere the equation

$$L(t, \hat{x}(t), \hat{u}(t)) + \langle \widehat{V}_x(t, \hat{x}(t)), f(t, \hat{x}(t), \hat{u}(t)) \rangle$$
$$= \min_{u \in U} \left\{ L(t, \hat{x}(t), u) + \langle \widehat{V}_x(t, \hat{x}(t)), f(t, \hat{x}(t), u) \rangle \right\} \tag{5.22}$$

which is equivalent to the Hamiltonian maximization condition

$$H(t, \hat{x}(t), \hat{u}(t), -\widehat{V}_x(t, \hat{x}(t))) = \max_{u \in U} H(t, \hat{x}(t), u, -\widehat{V}_x(t, \hat{x}(t))).$$

Then $\widehat{V}(t_0, x_0)$ is the optimal cost (i.e., $\widehat{V}(t_0, x_0) = V(t_0, x_0)$ where V is the value function) and $\hat{u}$ is an optimal control. (Note that this optimal control is not claimed to be unique; there can be multiple controls giving the same cost.)

To prove this result, let us first apply (5.20) with $x = \hat{x}(t)$. We know from (5.22) that along $\hat{x}$, the infimum is a minimum and it is achieved at $\hat{u}$; hence we have, similarly to (5.13),

$$-\widehat{V}_t(t, \hat{x}(t)) = L(t, \hat{x}(t), \hat{u}(t)) + \langle \widehat{V}_x(t, \hat{x}(t)), f(t, \hat{x}(t), \hat{u}(t)) \rangle.$$

We can move the $\widehat{V}_t$ term to the right-hand side and note that together with the inner product of $\widehat{V}_x$ and f it forms the total time derivative of $\widehat{V}$ along $\hat{x}$:

$$0 = L(t, \hat{x}(t), \hat{u}(t)) + \frac{d}{dt} \widehat{V}(t, \hat{x}(t)).$$

Integrating this equality with respect to t from t_0 to t_1, we have

$$0 = \int_{t_0}^{t_1} L(t, \hat{x}(t), \hat{u}(t)) dt + \widehat{V}(t_1, \hat{x}(t_1)) - \widehat{V}(t_0, \hat{x}(t_0))$$

which, in view of the boundary condition for $\widehat{V}$ and the initial condition for $\hat{x}$, gives

$$\widehat{V}(t_0, x_0) = \int_{t_0}^{t_1} L(t, \hat{x}(t), \hat{u}(t)) dt + K(\hat{x}(t_1)) = J(t_0, x_0, \hat{u}). \qquad (5.23)$$

On the other hand, if x is another trajectory with the same initial condition corresponding to an arbitrary control u, then (5.20) implies that

$$-\widehat{V}_t(t, x(t)) \leq L(t, x(t), u(t)) + \langle \widehat{V}_x(t, x(t)), f(t, x(t), u(t)) \rangle$$

or

$$0 \leq L(t, x(t), u(t)) + \frac{d}{dt} \widehat{V}(t, x(t)).$$

Integrating over $[t_0, t_1]$ as before, we obtain

$$0 \leq \int_{t_0}^{t_1} L(t, x(t), u(t)) dt + \widehat{V}(t_1, x(t_1)) - \widehat{V}(t_0, x(t_0))$$

or

$$\widehat{V}(t_0, x_0) \leq \int_{t_0}^{t_1} L(t, x(t), u(t)) dt + K(x(t_1)) = J(t_0, x_0, u). \qquad (5.24)$$

The equation (5.23) and the inequality (5.24) show that $\hat{u}$ gives the cost $\widehat{V}(t_0, x_0)$ while no other control u can produce a smaller cost. Thus we have confirmed that $\widehat{V}$ is the optimal cost and $\hat{u}$ is an optimal control.

We can regard the function $\widehat{V}$ in the above sufficient condition as providing a tool for verifying optimality of candidate optimal controls (obtained, for example, from the maximum principle). This optimality is automatically global. A simple modification of the above argument yields that $\widehat{V}(t, \hat{x}(t))$ is the optimal cost-to-go from an arbitrary point $(t, \hat{x}(t))$ on the trajectory $\hat{x}$. More generally, since $\widehat{V}$ is defined for all t and x, we could use an arbitrary pair (t, x) in place of (t_0, x_0) and obtain optimality with respect to $x(t) = x$ as the initial condition in the same way. Thus, if we have a family of controls parameterized by (t, x), each fulfilling the Hamiltonian maximization condition along the corresponding trajectory which starts at $x(t) = x$, then $\widehat{V}$ is the value function and it lets us establish optimality of all these controls. A typical way in which such a control family can arise is from a state feedback law description; we will encounter a scenario of this kind in Chapter 6. The next two exercises offer somewhat different twists on the above sufficient condition for optimality.

Exercise 5.3   *Suppose that a control $\hat{u} : [t_0, t_1] \to U$, the corresponding trajectory $\hat{x} : [t_0, t_1] \to \mathbb{R}^n$ with $\hat{x}(t_0) = x_0$, and a function $\widehat{V} : [t_0, t_1] \times \mathbb{R}^n \to \mathbb{R}$ are such that $\widehat{V}(t_0, x_0) = J(t_0, x_0, \hat{u})$ with J as in (5.1), i.e., $\widehat{V}(t_0, x_0)$ is the cost corresponding to the control $\hat{u}$. Suppose that $\widehat{V}$ satisfies the HJB equation (5.20) and the boundary condition (5.21). Show that then $\widehat{V}(t_0, x_0)$ is the optimal cost and $\hat{u}$ is an optimal control.* $\square$

Exercise 5.4   *Formulate and prove a sufficient condition for optimality, analogous to the one proved above, for the infinite-horizon problem described at the end of Section 5.1.3.* $\square$

5.1.5   Historical remarks

The HJB partial differential equation has its origins in the work of Hamilton, with subsequent improvements by Jacobi, done in the context of calculus of variations in the late 1830s. At that time the equation served as a necessary condition for optimality. Its use as a sufficient condition—still in the calculus of variations setting—was proposed in the work of Carathéodory, begun in the 1920s and culminating in his book published in 1935. (He established local optimality by working in a neighborhood of a test curve.) Carathéodory's approach became known as the "royal road" of calculus of variations.

The principle of optimality seems, in hindsight, an almost trivial observation, which actually dates all the way back to Jacob Bernoulli's 1697 solution of the brachistochrone problem. In the early 1950s, slightly before Bellman, the principle of optimality was formalized in the context of differential games by Isaacs, who called it the "tenet of transition." (The fundamental PDE of game theory bears Isaacs's initial alongside those of Hamilton, Jacobi, and Bellman.) The term "dynamic programming" was coined by Bellman, who published a series of papers and the book [Bel57] on this subject in the 1950s. Bellman's contribution was to recognize the power of the method to study value functions globally, and to use it for solving a variety of calculus of variations and optimal control problems.

It is not clear if Bellman was aware of the close connection between his work and the Hamilton-Jacobi equation of calculus of variations. This connection was explicitly made in the early 1960s by Kalman, who was apparently the first to use the name "HJB equation." Kalman's derivation of sufficient conditions for optimal control, combining the ideas of Carathéodory and Bellman, provided the basis for the treatment given here and in other modern sources. (The work of Kalman will also be prominently featured in the next chapter, where we discuss linear systems and quadratic costs.)

Quite remarkably, the maximum principle was being developed in the Soviet Union independently around the same time as Bellman's and Kalman's work on dynamic programming was appearing in the United States. We thus find it natural at this point to compare the maximum principle with the HJB equation and discuss the relationship between the two approaches.

5.2   HJB EQUATION VERSUS THE MAXIMUM PRINCIPLE

Here we focus on the *necessary* conditions for optimality provided by the HJB equation (5.10) and the Hamiltonian maximization condition (5.14) on one hand and by the maximum principle on the other hand. There is a notable difference in how these two necessary conditions characterize optimal controls. In order to see this point more clearly, assume that the system and the cost are time-invariant. The maximum principle is formulated in terms of the canonical equations

$$\dot{x}^* = H_p|_*, \qquad \dot{p}^* = -H_x|_* \qquad (5.25)$$

and says that at each time t, the value $u^*(t)$ of the optimal control must maximize $H(x^*(t), u, p^*(t))$ with respect to u:

$$u^*(t) = \arg \max_{u \in U} H(x^*(t), u, p^*(t)). \qquad (5.26)$$

This is an *open-loop* specification, because $u^*(t)$ depends not only on the state $x^*(t)$ but also on the costate $p^*(t)$ which has to be computed from the adjoint differential equation. Now, in the context of the HJB equation, the optimal control must satisfy

$$u^*(t) = \arg \max_{u \in U} H(x^*(t), u, -V_x(t, x^*(t))). \qquad (5.27)$$

This is a *closed-loop (feedback)* specification; indeed, assuming that we know the value function V everywhere, $u^*(t)$ is completely determined by the current state $x^*(t)$. The ability to generate an optimal control policy in the form of a state feedback law is an important feature of the dynamic programming approach, as we in fact already knew from Section 5.1.1. Clearly, we cannot implement this feedback law unless we can first find the value function by solving the HJB partial differential equation, and we have seen that this is in general a very difficult task. Therefore, from the computational point of view the maximum principle has an advantage in that it involves only ordinary and not partial differential equations. In principle, the dynamic programming approach provides more information (including sufficiency), but in reality, the maximum principle is often easier to use and allows one to solve many optimal control problems for which the HJB equation is analytically intractable.

As another point of comparison, it is interesting to recall how much longer and more complicated our proof of the maximum principle was compared with our derivation of the necessary conditions based on the HJB equation. This difference is especially perplexing in view of the striking similarity between the two Hamiltonian maximization conditions (5.26) and (5.27). We may wonder whether it might actually be possible to give an easier proof of the maximum principle starting from the HJB equation. Suppose that u^* is an optimal control and x^* is the corresponding state trajectory. Still assuming for simplicity that f and L are time-independent, we know that (5.27) must hold, where V is the value function satisfying

$$-V_t(t, x^*(t)) = L(x^*(t), u^*(t)) + \langle V_x(t, x^*(t)), f(x^*(t), u^*(t)) \rangle. \qquad (5.28)$$

To establish the maximum principle, we need to prove the existence of a costate p^* with the required properties. The formulas (5.26) and (5.27) strongly suggest that we should try to define it via

$$p^*(t) := -V_x(t, x^*(t)). \qquad (5.29)$$

Then, the desired Hamiltonian maximization condition (5.26) automatically follows from (5.27). We note also that if V satisfies the boundary condition $V(t_1, x) = K(x)$ as in (5.3), then the boundary condition for the costate (5.29) is $p^*(t_1) = -K_x(x^*(t_1))$, and this matches the boundary condition (4.43) that we had in the maximum principle for problems with terminal cost. Thus far, the situation looks quite promising, but we do not have any apparent reason to expect that p^* defined by (5.29) will satisfy the second differential equation in (5.25). However, this turns out to be true as well!

Exercise 5.5   *Let p^* be defined by (5.29), with V fulfilling (5.28). Prove that p^* satisfies the second canonical equation $\dot{p}^* = -H_x(x^*, u^*, p^*)$, where as usual $H(x, u, p) = \langle p, f(x, u) \rangle - L(x, u)$.* $\square$

In the proof of the maximum principle, the adjoint vector p^* was defined as the normal to a suitable hyperplane. In our earlier discussions in Section 3.4, it was also related to the momentum and to the vector of Lagrange multipliers. From (5.29) we now have another interpretation of the adjoint vector in terms of the gradient of the value function, i.e., the *sensitivity* of the optimal cost with respect to the state x. In economic terms, this quantity corresponds to the "marginal value," or "shadow price"; it tells us by how much we can increase benefits by increasing resources/spending, or how much we would be willing to pay someone else for resources and still make a profit.

At this point, the reader may be puzzled as to why we cannot indeed deduce the maximum principle from the HJB equation via the reasoning just given. Upon careful inspection, however, we can identify one gap in the above argument: it assumes that the value function has a well-defined gradient and, moreover, that this gradient can be further differentiated with respect to time (to obtain the adjoint equation as in Exercise 5.5). In other words, we need the existence of second-order partial derivatives of V. At the very least, we need V to be a C^1 function—a property that we have in fact assumed all along, starting with the Taylor expansion (5.7). The next example demonstrates that, unfortunately, we cannot expect this to be true in general.

5.2.1   Example: nondifferentiable value function

For the scalar system

$$\dot{x} = xu$$

with $x \in \mathbb{R}$ and $u \in [-1, 1]$, consider a fixed-time, free-endpoint problem with the cost $J(u) = x(t_1)$. The optimal solution is easily found by inspection: if $x_0 > 0$ then apply $u \equiv -1$ which results in $\dot{x} = -x$, hence the cost is $x(t_1) = e^{-(t_1-t_0)}x_0$; if $x_0 < 0$ then use $u \equiv 1$ which gives $\dot{x} = x$ and the cost is $x(t_1) = e^{t_1-t_0}x_0$; finally, if $x_0 = 0$ then $x \equiv 0$ for all u and the cost is 0. We see that the value function is given by

$$V(t, x) = \begin{cases} e^{-(t_1-t)}x & \text{if } x > 0, \\ e^{t_1-t}x & \text{if } x < 0, \\ 0 & \text{if } x = 0. \end{cases} \tag{5.30}$$

Away from $x = 0$ it indeed satisfies the HJB equation for this example, which is $-V_t = \inf_u \{V_x\, xu\} = -|V_x\, x|$, with the boundary condition $V(t_1, x) = x$. For a fixed $t < t_1$, the graph of V as a function of x is plotted in Figure 5.4. At $x = 0$ this function is Lipschitz but not C^1. It can actually be shown that the above HJB equation does not admit any C^1 solution.

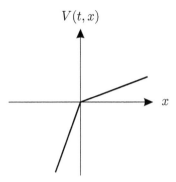

Figure 5.4: Value function nondifferentiable at $x = 0$

It turns out that this state of affairs is not an exception; in fact, it is quite typical for problems with bounded controls and terminal cost to have nondifferentiable value functions. On the other hand, the local Lipschitz property—which the function (5.30) does possess—is a known attribute of value functions for some reasonably general classes of optimal control problems (we will say more on this below).

The above example clarifies why we cannot derive the maximum principle from the HJB equation. There really is no "easy" proof of the maximum principle (except in settings much less general than the one we considered). More importantly, the difficulty that we just exposed has implications not only for relating the HJB equation and the maximum principle, but for the HJB theory itself. Namely, we need to reconsider the assumption that $V \in \mathcal{C}^1$ and instead work with some generalized concept of a solution to the HJB partial differential equation.[3] Because of this difficulty, the theory of dynamic programming did not become rigorous until the early 1980s when, after a series of related developments, the notion of a *viscosity solution* was introduced by Crandall and Lions; that work completes the historical timeline of key contributions listed in Section 5.1.5. (The maximum principle, on the other hand, was on solid technical ground from the beginning.) We turn to viscosity solutions in the next section, postponing a discussion of further links between the HJB equation and the maximum principle until Section 7.2.

---

[3] We note that generalized solution concepts, particularly those relaxing the continuous differentiability requirement, are important in the theory of ordinary differential equations as well. One such solution concept (albeit a very simple one) is provided by the class of absolutely continuous functions, as we discussed in Section 3.3.1.

5.3   VISCOSITY SOLUTIONS OF THE HJB EQUATION

We first need to familiarize ourselves with a few basic notions and results from *nonsmooth analysis*.

5.3.1   One-sided differentials

Let $v : \mathbb{R}^n \to \mathbb{R}$ be a continuous function (nothing beyond continuity is required from v). A vector $\xi \in \mathbb{R}^n$ is called a *super-differential* of v at a given point x if for all y near x we have the relation

$$v(y) \leq v(x) + \langle \xi, y - x \rangle + o(|y - x|). \tag{5.31}$$

Geometrically, ξ is a super-differential if the graph of the linear function $y \mapsto v(x) + \langle \xi, y - x \rangle$, which has ξ as its gradient and takes the value $v(x)$ at $y = x$, lies above the graph of v at least locally near x (or is tangent to the graph of v at x). Figure 5.5 (a) illustrates this situation for the scalar case $(n = 1)$, in which ξ is the slope of the line. A super-differential ξ is in general not unique; we thus have a *set* of super-differentials of v at x, which is denoted by $D^+v(x)$.

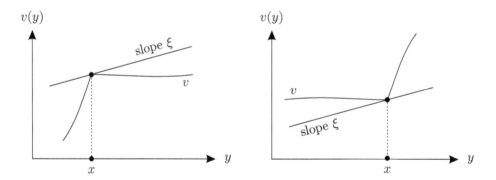

Figure 5.5: (a) super-differential, (b) sub-differential

Similarly, we say that $\xi \in \mathbb{R}^n$ is a *sub-differential* of v at x if

$$v(y) \geq v(x) + \langle \xi, y - x \rangle - o(|y - x|). \tag{5.32}$$

The graph of the linear function with gradient ξ touching the graph of v at x must now lie *below* the graph of v in a vicinity of x (or be tangent to it at x); see Figure 5.5 (b). The set of sub-differentials of v at x is denoted by $D^-v(x)$.

Example 5.3   *For the function*

$$v(x) = \begin{cases} 0 & \text{if } x < 0, \\ \sqrt{x} & \text{if } 0 \le x \le 1, \\ 1 & \text{if } x > 1 \end{cases}$$

plotted in Figure 5.6, the reader should have no difficulty in verifying that $D^+v(0) = \emptyset$, $D^-v(0) = [0, \infty)$, $D^+v(1) = [0, 1/2]$, *and* $D^-v(1) = \emptyset$. *As we will see shortly, these points at which v is not differentiable are the only "interesting" points to check.* □

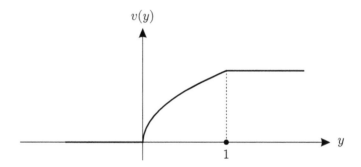

Figure 5.6: The function in Example 5.3

We now establish some useful properties of super- and sub-differentials.

TEST FUNCTIONS. $\xi \in D^+v(x)$ if and only if there exists a $\mathcal{C}^1$ function $\varphi : \mathbb{R}^n \to \mathbb{R}$ such that $\nabla\varphi(x) = \xi$, $\varphi(x) = v(x)$, and $\varphi(y) \ge v(y)$ for all y near x, i.e., $\varphi - v$ has a local minimum at x. Similarly, $\xi \in D^-v(x)$ if and only if there exists a $\mathcal{C}^1$ function φ such that $\nabla\varphi(x) = \xi$ and $\varphi - v$ has a local *maximum* at x. (Note that we can always arrange to have $\varphi(x) = v(x)$ by adding a constant to φ, which does not affect the other conditions.)

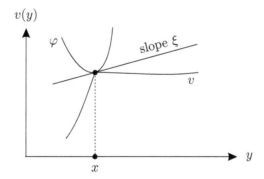

Figure 5.7: Characterization of a super-differential via a test function

The function φ is sometimes called a *test function*. For the case of $D^+v(x)$, an example of such a function is shown in Figure 5.7. The above result, whose proof we will not give, will be used for proving the other facts that follow.

RELATION WITH CLASSICAL DIFFERENTIALS. If v is differentiable at x, then

$$D^+v(x) = D^-v(x) = \{\nabla v(x)\}. \tag{5.33}$$

If $D^+v(x)$ and $D^-v(x)$ are both nonempty, then v is differentiable at x and the relation (5.33) holds.

We prove both claims with the help of test functions. First, suppose that v is differentiable at x. It is clear that the gradient $\nabla v(x)$ is both a super-differential and a sub-differential of v at x (indeed, with $\xi = \nabla v(x)$ both (5.31) and (5.32) become equalities). If $\varphi - v$ has a local minimum or maximum at x for some $\varphi \in \mathcal{C}^1$, then $\nabla(\varphi - v)(x) = 0$ hence $\nabla\varphi(x) = \nabla v(x)$. This shows that $\xi = \nabla v(x)$ is the only element in $D^+v(x)$ and $D^-v(x)$.

To prove the second claim, let $\varphi_1, \varphi_2 \in \mathcal{C}^1$ be such that $\varphi_1(x) = v(x) = \varphi_2(x)$ and $\varphi_1(y) \le v(y) \le \varphi_2(y)$ for y near x. Then $\varphi_1 - \varphi_2$ has a local maximum at x, implying that $\nabla(\varphi_1 - \varphi_2)(x) = 0$ hence $\nabla\varphi_1(x) = \nabla\varphi_2(x)$. For $y > x$ (the case $y < x$ is completely analogous) we can write

$$\frac{\varphi_1(y) - \varphi_1(x)}{y - x} \le \frac{v(y) - v(x)}{y - x} \le \frac{\varphi_2(y) - \varphi_2(x)}{y - x}.$$

As $y \to x$, the first fraction approaches $\nabla\varphi_1(x)$, the last one approaches $\nabla\varphi_2(x)$, and we know that the two gradients are equal. Therefore, by the "Sandwich Theorem" the limit of the middle fraction exists and equals the other two. This limit must be $\nabla v(x)$, and everything is proved.

NON-EMPTINESS AND DENSENESS. The sets $\{x : D^+v(x) \ne \emptyset\}$ and $\{x : D^-v(x) \ne \emptyset\}$ are both nonempty, and actually dense in the domain of v.

The idea of the proof is sketched in Figure 5.8. The highly irregular graph in the figure is that of v. Take an arbitrary point x_0. Choosing a $\mathcal{C}^1$ function φ steep enough (see the solid curve in the figure), we can force $\varphi - v$ to have a local maximum at a nearby point x, as close to x_0 as we want. (For clarity, we also shift the graph of φ vertically to produce the dotted curve in the figure which touches the graph of v at x.) We then have $\nabla\varphi(x) \in D^-v(x)$. A similar argument works for D^+.

Exercise 5.6 *Give an example of a continuous function v with the property that for some x in its domain both $D^+v(x)$ and $D^-v(x)$ are empty.* $\square$

5.3.2 Viscosity solutions of PDEs

We are now ready to introduce the concept of a viscosity solution for PDEs. Consider a PDE of the form

$$F(x, v(x), \nabla v(x)) = 0 \tag{5.34}$$

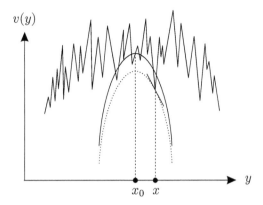

Figure 5.8: Proving denseness

where $F : \mathbb{R}^n \times \mathbb{R} \times \mathbb{R}^n \to \mathbb{R}$ is a continuous function. A *viscosity subsolution* of the PDE (5.34) is a continuous function $v : \mathbb{R}^n \to \mathbb{R}$ such that

$$F(x, v(x), \xi) \leq 0 \qquad \forall \, \xi \in D^+ v(x), \ \forall \, x. \qquad (5.35)$$

As we know, this is equivalent to saying that at every x we must have $F(x, v(x), \nabla\varphi(x)) \leq 0$ for every $\mathcal{C}^1$ test function φ such that $\varphi - v$ has a local minimum at x. Similarly, v is a *viscosity supersolution* of (5.34) if

$$F(x, v(x), \xi) \geq 0 \qquad \forall \, \xi \in D^- v(x), \ \forall \, x \qquad (5.36)$$

or, equivalently, at every x we have $F(x, v(x), \nabla\varphi(x)) \geq 0$ for every $\mathcal{C}^1$ function φ such that $\varphi - v$ has a local maximum at x. Finally, v is a *viscosity solution* if it is both a viscosity subsolution and a viscosity supersolution.

The above definitions of a viscosity subsolution and supersolution impose conditions on v only at points where $D^+ v$, respectively $D^- v$, is nonempty. We know that the set of these points is dense in the domain of v. At all points where v is differentiable, the PDE must hold in the classical sense. If v is Lipschitz, then by Rademacher's theorem it is differentiable almost everywhere.

Example 5.4 *Consider the scalar case ($n = 1$) and let $F(x, v, \xi) = 1 - |\xi|$, for which the PDE (5.34) is $1 - |\nabla v(x)| = 0$. The functions $v(x) = x$ and $v(x) = -x$ are both classical solutions of this PDE (so are the functions $v(x) = \pm x + c$ where c is a constant). We claim that the function $v(x) = |x|$ is a viscosity solution. For all $x \neq 0$ this v is differentiable and the PDE is satisfied, thus we only need to check what happens at $x = 0$. First, $D^+ v(0) = \emptyset$ hence (5.35) is true. Second, $D^- v(0) = [-1, 1]$ and the inequality $1 - |\xi| \geq 0$ holds for all $\xi \in [-1, 1]$, making (5.36) true as well.* $\square$

Note the lack of symmetry in the definition of a viscosity solution: the sign convention used when writing the PDE is important. In the above example, if we rewrite the PDE as $|\nabla v(x)| - 1 = 0$, then it is easy to see that $v(x) = |x|$ is no longer a viscosity solution.

The terminology "viscosity solutions" is motivated by the fact that a viscosity solution v of the PDE (5.34) can be obtained from smooth solutions v_ε to the family of PDEs

$$F(x, v_\varepsilon(x), \nabla v_\varepsilon(x)) = \varepsilon \Delta v_\varepsilon(x) \tag{5.37}$$

(parameterized by $\varepsilon > 0$) in the limit as $\varepsilon \to 0$. The operator Δ on the right-hand side of (5.37) denotes the Laplacian ($\Delta v = v_{x_1 x_1} + \cdots + v_{x_n x_n}$); in fluid mechanics it appears in the PDE describing the motion of a viscous fluid. To understand the basic idea behind this convergence result, let $\xi \in D^- v(x_0)$ for some x_0. Consider a test function φ such that $\nabla \varphi(x_0) = \xi$ and $\varphi - v$ has a local maximum at x_0. Assume that $\varphi \in \mathcal{C}^2$ (if not, approximate it by a $\mathcal{C}^2$ function). If v_ε is close to v for small ε, then $\varphi - v_\varepsilon$ has a local maximum at some x_ε near x_0, implying that $\nabla \varphi(x_\varepsilon) = \nabla v_\varepsilon(x_\varepsilon)$ and $\Delta \varphi(x_\varepsilon) \leq \Delta v_\varepsilon(x_\varepsilon)$. Since v_ε solves (5.37), this gives $F(x_\varepsilon, v_\varepsilon(x_\varepsilon), \nabla \varphi(x_\varepsilon)) \geq \varepsilon \Delta \varphi(x_\varepsilon)$. Taking the limit as $\varepsilon \to 0$, by continuity of F we have $F(x_0, v(x_0), \nabla \varphi(x_0)) \geq 0$, which means that v is a supersolution of (5.34). The argument showing that v is a subsolution is similar.

5.3.3   HJB equation and the value function

Let us finally go back to our fixed-time optimal control problem from Section 5.1.2 and its HJB equation (5.10) with the boundary condition (5.3), with the goal of resolving the difficulty identified at the end of Section 5.2.1. Specifically, our hope is that the value function (5.2) is a solution of the HJB equation in the viscosity sense. In order to more closely match the PDE (5.34), we first rewrite the HJB equation as

$$-V_t(t, x) - \inf_{u \in U} \{ L(t, x, u) + \langle V_x(t, x), f(t, x, u) \rangle \} = 0. \tag{5.38}$$

As we saw in the previous subsection, flipping the sign in a PDE affects its viscosity solutions; it will turn out that the above sign convention is the correct one. The PDE (5.38) is still not in the form (5.34) because it contains the additional independent variable t. However, the concepts and results of Sections 5.3.1 and 5.3.2 extend to this time-space setting without difficulties. Alternatively, we can absorb t into x by introducing the extra state variable $x_{n+1} := t$, a trick already familiar to us from Chapters 3 and 4. Then the PDE (5.38) takes the form (5.34) with the obvious definition of the function F, except that the domain of v is the subset[4] $[t_0, t_1] \times \mathbb{R}^n$ of $\mathbb{R}^{n+1}$.

---

[4] In Sections 5.3.1 and 5.3.2 we could have easily taken the domain of v to be a subset of $\mathbb{R}^n$ rather than the entire $\mathbb{R}^n$.

The control u does not appear as an argument of F since the definition of F includes taking the infimum over u.

Now the theory of viscosity solutions can be applied to the HJB equation. Under suitable technical assumptions on the functions f, L, and K, we have the following **main result**: *The value function V is a unique viscosity solution of the HJB equation* (5.38) *with the boundary condition* (5.3). *It is also locally Lipschitz* (but, as we know from Section 5.2.1, not necessarily C^1). Regarding the technical assumptions, we will not list them here (see the references in Section 5.4) but we mention that they are satisfied if, for example, f, L, and K are uniformly continuous, f_x, L_x, and K_x are bounded, and U is a compact set.

We will not attempt to establish the uniqueness and the Lipschitz property, but we do want to understand why V is a viscosity solution of (5.38). To this end, let us prove that V is a viscosity *sub*solution. The additional technical assumptions cited above are actually not needed for this claim. Fix an arbitrary pair (t_0, x_0). We need to show that for every C^1 test function $\varphi = \varphi(t, x)$ such that $\varphi - V$ attains a local minimum at (t_0, x_0), the inequality

$$-\varphi_t(t_0, x_0) - \inf_{u \in U} \{L(t_0, x_0, u) + \langle \varphi_x(t_0, x_0), f(t_0, x_0, u) \rangle\} \leq 0$$

must be satisfied. Suppose that, on the contrary, there exist a C^1 function φ and a control value $u_0 \in U$ such that

$$\varphi(t_0, x_0) = V(t_0, x_0), \qquad \varphi(t, x) \geq V(t, x) \quad \forall\, (t, x) \text{ near } (t_0, x_0) \quad (5.39)$$

and

$$-\varphi_t(t_0, x_0) - L(t_0, x_0, u_0) - \langle \varphi_x(t_0, x_0), f(t_0, x_0, u_0) \rangle > 0. \qquad (5.40)$$

Taking (t_0, x_0) as the initial condition, let us consider the state trajectory $x(\cdot)$ that results from applying the control $u \equiv u_0$ on a small time interval $[t_0, t_0 + \Delta t]$. We will now demonstrate that the rate of change of the value function along this trajectory is inconsistent with the principle of optimality. As long as we pick Δt to be sufficiently small, we have

$$V(t_0 + \Delta t, x(t_0 + \Delta t)) - V(t_0, x_0) \leq \varphi(t_0 + \Delta t, x(t_0 + \Delta t)) - \varphi(t_0, x_0)$$

$$= \int_{t_0}^{t_0 + \Delta t} \frac{d}{dt} \varphi(t, x(t)) dt = \int_{t_0}^{t_0 + \Delta t} \left(\varphi_t(t, x(t)) + \langle \varphi_x(t, x(t)), f(t, x(t), u_0) \rangle \right) dt$$

$$< - \int_{t_0}^{t_0 + \Delta t} L(t, x(t), u_0) dt$$

where the first inequality is a direct consequence of (5.39) and the last inequality follows from (5.40) by virtue of continuity of all the functions appearing there. We thus obtain

$$V(t_0, x_0) > \int_{t_0}^{t_0+\Delta t} L(t, x(t), u_0)dt + V(t_0 + \Delta t, x(t_0 + \Delta t)). \qquad (5.41)$$

On the other hand, the principle of optimality (5.4) tells us that

$$V(t_0, x_0) \leq \int_{t_0}^{t_0+\Delta t} L(t, x(t), u_0)dt + V(t_0 + \Delta t, x(t_0 + \Delta t))$$

and we arrive at a contradiction. Provided that optimal controls exist, here is a slightly different way to see why (5.41) cannot be true: it would imply that the optimal cost-to-go from (t_0, x_0) is higher than the cost of applying the constant control $u \equiv u_0$ on $[t_0, t_0 + \Delta t]$ followed by an optimal control on the remaining interval $(t_0 + \Delta t, t_1]$, which is clearly impossible.

Exercise 5.7 *Use a similar argument to show that the value function is a viscosity supersolution of the HJB equation (5.38).* $\square$

We now have at our disposal a more rigorous formulation of the necessary conditions for optimality from Section 5.1.3. The above reasoning is of course quite different from our original derivation of the HJB equation, but we see that the principle of optimality still plays a central role. The sufficient condition for optimality from Section 5.1.4 can also be generalized, a task that we leave as an exercise.

Exercise 5.8 *Formulate and prove a sufficient condition for optimality analogous to that of Exercise 5.3 in the framework of viscosity solutions.* $\square$

5.4   NOTES AND REFERENCES FOR CHAPTER 5

The treatment of the discrete problem in Section 5.1.1 is based on [Son98, Section 8.1]. The material of Sections 5.1.2–5.1.4 is quite standard; it is assembled from the texts [AF66, BP07, LM67, Son98, YZ99] which further elaborate on some aspects of this theory. The brief historical remarks of Section 5.1.5 are compiled largely from [YZ99]; for additional reading on how the subject has evolved we recommend [Bry96] and [PB94] (the latter paper is more technical and explains how Carathéodory's work was, along with subsequent work of Hestenes, also a precursor of the maximum principle). A derivation of the Hamilton-Jacobi PDE as a necessary condition for optimality in the context of calculus of variations (via the general formula for the variation of a functional), as well as an explanation of its connection to Hamilton's canonical equations, can be found in [GF63, Section 23]; see also [YZ99, pp. 222–223].

Our discussion of the relationship between the HJB equation and the maximum principle in Section 5.2 follows [YZ99, pp. 229–230] and [Vin00, Section 1.6]. The economic interpretation of the adjoint vector and the value function is developed in more detail in [YZ99, pp. 231–232]; similar ideas appear in [Lue84, Section 4.4] and are taken further in [Cla10]. The example of Section 5.2.1 is Example 2.3 in [YZ99, Chapter 4], where it is proved that the corresponding HJB equation admits no $\mathcal{C}^1$ solution. The same example is also studied in [Vin00, Section 1.7], and a different example with similar features is Example 7.2 in [BP07]. For results showing—under different assumptions—that the value function is locally Lipschitz, see Theorem 2.5 in [YZ99, Chapter 4], Lemma 8.6.2 in [BP07], or Proposition 12.3.5 in [Vin00].

Section 5.3 is heavily based on the exposition in [BP07, Chapter 8]. The original reference on viscosity solutions is [CL83]. The convergence result sketched at the end of Section 5.3.2 is Theorem 8.4.2 in [BP07]; for background on the motion of viscous fluids the reader can consult [FLS63, Chapter II-41]. A complete proof of the main result in Section 5.3.3, again under different technical assumptions, can be found in [BP07] and [YZ99]. Other generalized solution concepts for HJB equations have been proposed; see the notes and references in [YZ99, Chapter 4] and [Vin00, Chapter 12]. For an in-depth study of HJB equations and their viscosity solutions, including numerical methods, see [BCD97].

Chapter Six

The Linear Quadratic Regulator

6.1 FINITE-HORIZON LQR PROBLEM

In this chapter we will focus on the special case when the system dynamics are linear and the cost is quadratic. While this additional structure certainly makes the optimal control problem more tractable, our goal is not merely to specialize our earlier results to this simpler setting. Rather, we want to go deeper and develop a more complete understanding of optimal solutions compared with what we were able to achieve for the general scenarios treated in the previous chapters. (We could have followed a different path in our studies and started with this specific problem class before tackling more difficult problems. From the pedagogical point of view, each approach has its own merits. Historically, however, the general nonlinear results—having their origins in calculus of variations—appeared first.)

The (finite-horizon) *Linear Quadratic Regulator (LQR)* problem is the optimal control problem from Section 3.3 with the following additional assumptions: the control system is a linear time-varying system

$$\dot{x} = A(t)x + B(t)u, \qquad x(t_0) = x_0 \tag{6.1}$$

with $x \in \mathbb{R}^n$ and $u \in \mathbb{R}^m$ (the control is unconstrained); the target set is $S = \{t_1\} \times \mathbb{R}^n$, where t_1 is a fixed time (so this is a fixed-time, free-endpoint problem); and the cost functional is

$$J(u) = \int_{t_0}^{t_1} \left(x^T(t)Q(t)x(t) + u^T(t)R(t)u(t) \right) dt + x^T(t_1)Mx(t_1) \tag{6.2}$$

where $Q(\cdot)$, $R(\cdot)$, M are matrices of appropriate dimensions satisfying $M = M^T \geq 0$ (symmetric positive semidefinite), $Q(t) = Q^T(t) \geq 0$ (symmetric positive semidefinite), and $R(t) = R^T(t) > 0$ (symmetric positive *definite*) for all $t \in [t_0, t_1]$. The quadratic cost (6.2) is very reasonable: since both Q and R are positive (semi)definite, it penalizes both the size of the state and the control effort, with Q and R determining their relative weights. Incidentally, the formula $L(t, x, u) = x^T Q(t)x + u^T R(t)u$ for the running cost is another justification of the acronym LQR. We require R to be strictly positive definite because we will soon need its inverse. Additional assumptions will be introduced later as necessary. More general target sets can also be considered.

6.1.1 Candidate optimal feedback law

We begin our analysis of the LQR problem by inspecting the necessary conditions for optimality provided by the maximum principle. After some further manipulation, these conditions will reveal that an optimal control must be a linear state feedback law. The Hamiltonian is given by

$$H(t, x, u, p) = p^T A(t)x + p^T B(t)u - x^T Q(t)x - u^T R(t)u.$$

Note that, compared to the general formula (4.40) for the Hamiltonian, we took the abnormal multiplier p_0 to be equal to -1. This is no loss of generality because, for the present free-endpoint problem in the Bolza form, a combination of the results in Section 4.3.1 would give us the transversality condition $0 = p^*(t_1) - p_0^* K_x(x^*(t_1)) = p^*(t_1) - 2p_0^* M x^*(t_1)$ which, in light of the nontriviality condition, guarantees that $p_0^* \neq 0$. It is also useful to observe that the LQR problem can be adequately treated with the variational approach of Section 3.4, which yields essentially the same necessary conditions as the maximum principle but without the abnormal multiplier appearing. Indeed, the control is unconstrained, the final state is free, and H is quadratic—hence twice differentiable—in u; therefore, the technical issues discussed in Section 3.4.5 do not arise here. In Section 3.4 we proved that along an optimal trajectory we must have $H_u|_* = 0$ and $H_{uu}|_* \leq 0$, which is in general different from the Hamiltonian maximization condition, but in the present LQR setting this difference disappears as we will see in a moment. In fact, when solving part a) of Exercise 3.8, the reader should have already written down the necessary conditions for optimality from Section 3.4.3 for the LQR problem (with $M = 0$). We will now rederive these necessary conditions and examine their consequences in more detail.

The gradient of H with respect to u is $H_u = B^T(t)p - 2R(t)u$, and along an optimal trajectory it must vanish. Using our assumption that $R(t)$ is invertible for all t, we can solve the resulting equation for u and conclude that an optimal control u^* (if it exists) must satisfy

$$u^*(t) = \frac{1}{2}R^{-1}(t)B^T(t)p^*(t). \tag{6.3}$$

Moreover, since $H_{uu} = -2R(t) < 0$, the above control indeed maximizes the Hamiltonian (globally). We see that (6.3) is the unique control satisfying the necessary conditions, although we have not yet verified that it is optimal.

Since the formula (6.3) expresses u^* in terms of the costate p^*, let us look at p^* more closely. It satisfies the adjoint equation

$$\dot{p}^* = -H_x|_* = 2Q(t)x^* - A^T(t)p^* \tag{6.4}$$

with the boundary condition

$$p^*(t_1) = -K_x(x^*(t_1)) = -2Mx^*(t_1) \tag{6.5}$$

(see Section **4.3.1**), where x^* is the optimal state trajectory. Our next goal is to show that a linear relation of the form

$$p^*(t) = -2P(t)x^*(t) \tag{6.6}$$

holds for all t and not just for $t = t_1$, where $P(\cdot)$ is some matrix-valued function to be determined. Putting together the dynamics (6.1) of the state, the control law (6.3), and the dynamics (6.4) of the costate, we can write the system of canonical equations in the following combined closed-loop form:

$$\begin{pmatrix} \dot{x}^* \\ \dot{p}^* \end{pmatrix} = \begin{pmatrix} A(t) & \frac{1}{2}B(t)R^{-1}(t)B^T(t) \\ 2Q(t) & -A^T(t) \end{pmatrix} \begin{pmatrix} x^* \\ p^* \end{pmatrix} =: \mathcal{H}(t) \begin{pmatrix} x^* \\ p^* \end{pmatrix}. \tag{6.7}$$

The matrix $\mathcal{H}(t)$ is sometimes called the *Hamiltonian matrix*. Let us denote the state transition matrix for the linear time-varying system (6.7) by $\Phi(\cdot, \cdot)$. Then we have, in particular, $\begin{pmatrix} x^*(t) \\ p^*(t) \end{pmatrix} = \Phi(t, t_1) \begin{pmatrix} x^*(t_1) \\ p^*(t_1) \end{pmatrix}$; here $\Phi(t, t_1) = \Phi^{-1}(t_1, t)$ propagates the solutions backward in time from t_1 to t. Partitioning Φ into $n \times n$ blocks as

$$\Phi = \begin{pmatrix} \Phi_{11} & \Phi_{12} \\ \Phi_{21} & \Phi_{22} \end{pmatrix}$$

gives the more detailed relation

$$\begin{pmatrix} x^*(t) \\ p^*(t) \end{pmatrix} = \begin{pmatrix} \Phi_{11}(t, t_1) & \Phi_{12}(t, t_1) \\ \Phi_{21}(t, t_1) & \Phi_{22}(t, t_1) \end{pmatrix} \begin{pmatrix} x^*(t_1) \\ p^*(t_1) \end{pmatrix}$$

which, in view of the terminal condition (6.5), can be written as

$$x^*(t) = (\Phi_{11}(t, t_1) - 2\Phi_{12}(t, t_1)M)x^*(t_1), \tag{6.8}$$
$$p^*(t) = (\Phi_{21}(t, t_1) - 2\Phi_{22}(t, t_1)M)x^*(t_1). \tag{6.9}$$

Solving (6.8) for $x^*(t_1)$ and plugging the result into (6.9), we obtain

$$p^*(t) = \big(\Phi_{21}(t, t_1) - 2\Phi_{22}(t, t_1)M\big)\big(\Phi_{11}(t, t_1) - 2\Phi_{12}(t, t_1)M\big)^{-1}x^*(t).$$

We have thus established (6.6) with

$$P(t) := -\frac{1}{2}\big(\Phi_{21}(t, t_1) - 2\Phi_{22}(t, t_1)M\big)\big(\Phi_{11}(t, t_1) - 2\Phi_{12}(t, t_1)M\big)^{-1}. \tag{6.10}$$

   A couple of remarks are in order. First, we have not yet justified the existence of the inverse in the definition of $P(t)$. For now, we note that

$\Phi(t_1, t_1) = I_{2n \times 2n}$, hence $\Phi_{11}(t_1, t_1) = I_{n \times n}$, $\Phi_{12}(t_1, t_1) = 0_{n \times n}$, and so $\Phi_{11}(t_1, t_1) - 2\Phi_{12}(t_1, t_1)M = I_{n \times n}$. By continuity, $\Phi_{11}(t, t_1) - 2\Phi_{12}(t, t_1)M$ stays invertible for t close enough to t_1, which means that $P(t)$ is well defined at least for t near t_1. Second, the minus sign and the factor of $1/2$ in (6.10), which stem from the factor of -2 in (6.6), appear to be somewhat arbitrary at this point. We see from (6.5) and (6.6) that

$$P(t_1) = M. \tag{6.11}$$

The reason for the above conventions will become clear later, when we show that $P(t)$ is symmetric positive semidefinite for all t (not just for $t = t_1$) and is directly related to the optimal cost.

Combining (6.3) and (6.6), we deduce that the optimal control must take the form

$$\boxed{u^*(t) = -R^{-1}(t)B^T(t)P(t)x^*(t)} \tag{6.12}$$

which, as we announced earlier, is a *linear state feedback law*. This is a remarkable conclusion, as it shows that the optimal closed-loop system must be linear.[1] Note that the feedback gain in (6.12) is time-varying even if the system (6.1) is time-invariant, because P from (6.10) is always time-varying. We remark that we could just as easily derive an open-loop formula for u^* by writing (6.8) with t_0 in place of t, i.e., $x_0 = (\Phi_{11}(t_0, t_1) - 2\Phi_{12}(t_0, t_1)M)x^*(t_1)$, solving it for $x^*(t_1)$ and plugging the result into (6.9) to arrive at

$$p^*(t) = \left(\Phi_{21}(t, t_1) - 2\Phi_{22}(t, t_1)M\right)\left(\Phi_{11}(t_0, t_1) - 2\Phi_{12}(t_0, t_1)M\right)^{-1}x_0$$

(provided that the inverse exists), and then using this expression in (6.3). However, the feedback form of u^* is theoretically revealing and leads to a more compact description of the closed-loop system.

As we said, there are two things that we still need to check: optimality of the control u^* that we found, and global existence of the matrix $P(t)$. These issues will be tackled in Sections 6.1.3 and 6.1.4, respectively. But first, we want to obtain a nicer description for the matrix $P(t)$, as the formula (6.10) is rather clumsy and not very useful (since calculating the state transition matrix Φ analytically is in general impossible).

6.1.2   Riccati differential equation

We will now derive a *differential equation* that the matrix $P(\cdot)$ defined in (6.10) must satisfy. First, differentiate both sides of the equality (6.6) to obtain

$$\dot{p}^*(t) = -2\dot{P}(t)x^*(t) - 2P(t)\dot{x}^*(t).$$

---

[1]Of course, the idea that quadratic optimization problems lead to linear update laws goes back to Gauss's least squares method.

Next, expand $\dot{p}^*$ and $\dot{x}^*$ using the canonical equations (6.7) to arrive at

$$2Q(t)x^*(t) - A^T(t)p^*(t)$$
$$= -2\dot{P}(t)x^*(t) - 2P(t)A(t)x^*(t) - P(t)B(t)R^{-1}(t)B^T(t)p^*(t).$$

Applying (6.6) to eliminate p^* and dividing by 2, we conclude that the equation

$$Q(t)x^*(t) + A^T(t)P(t)x^*(t)$$
$$= -\dot{P}(t)x^*(t) - P(t)A(t)x^*(t) + P(t)B(t)R^{-1}(t)B^T(t)P(t)x^*(t) \quad (6.13)$$

must hold (for all t at which $P(t)$ is defined). Since the initial state x_0 is arbitrary and x^* is the state of the linear time-varying system given by (6.1) and (6.12) whose state transition matrix is nonsingular, $x^*(t)$ can be arbitrary. It follows that P must be a solution of the matrix differential equation

$$\boxed{\dot{P}(t) = -P(t)A(t) - A^T(t)P(t) - Q(t) + P(t)B(t)R^{-1}(t)B^T(t)P(t)}$$
$$(6.14)$$

which is called the **Riccati differential equation (RDE)**. We already know that the boundary condition for it is specified by (6.11). The solution $P(t)$ is to be propagated backward in time from $t = t_1$; its global existence remains to be addressed.

It is interesting to compare the two descriptions that we now have for $P(t)$. The RDE (6.14) is a quadratic matrix differential equation. The formula (6.10), on the other hand, is in terms of the state transition matrix Φ which satisfies a *linear* matrix differential equation $\frac{d}{dt}\Phi(t, t_1) = \mathcal{H}(t)\Phi(t, t_1)$ but has size $2n \times 2n$ (while P is $n \times n$). Ignoring the computational effort involved in computing the matrix inverse in (6.10), we can say that by passing from (6.10) to (6.14) we reduced in half the size of the matrix to be solved for, but traded a linear differential equation for a quadratic one. Actually, if we prefer matrix differential equations that are linear rather than quadratic, it is possible to compute $P(t)$ somewhat more efficiently by solving a linear system of size $2n \times n$, as shown in the next exercise.

Exercise 6.1   *Let $X(t)$, $Y(t)$ be $n \times n$ matrices satisfying the linear differential equation*

$$\begin{pmatrix} \dot{X} \\ \dot{Y} \end{pmatrix} = \mathcal{H}(t) \begin{pmatrix} X \\ Y \end{pmatrix}$$

with the boundary condition $X(t_1) = I$, $Y(t_1) = -2M$, where $\mathcal{H}(t)$ is defined in (6.7). Check that the matrix

$$P(t) := -\frac{1}{2}Y(t)X^{-1}(t) \tag{6.15}$$

satisfies the RDE (6.14) *and the boundary condition* (6.11).   □

The idea of reducing a quadratic differential equation to a linear one of twice the size is in fact not new to us; we already saw it in Section 2.6.2 in the context of deriving second-order sufficient conditions for optimality in calculus of variations. In the single-degree-of-freedom case, we passed from the first-order quadratic differential equation (2.64) to the second-order linear differential equation (2.67) via the substitution (2.66). In the multiple-degrees-of-freedom setting, scalar variables need to be replaced by matrices but a similar transformation can be applied, as we stated (without including the derivations) at the end of Section 2.6.2. Associating the matrix W there with the matrix P here, the reader will readily see the correspondence between that earlier construction and the one given in Exercise 6.1.

The outcome of applying the necessary conditions of the maximum principle to the LQR problem can now be summarized as follows: a unique candidate for an optimal control is given by the linear feedback law (6.12), where the matrix $P(t)$ satisfies the RDE (6.14) and the boundary condition (6.11). This is as far as the maximum principle can take us; we need to employ other tools for investigating whether $P(t)$ exists for all t and whether the control (6.12) is indeed optimal.

6.1.3   Value function and optimality

We now proceed to show that the control law (6.12) identified by the maximum principle is globally optimal. We already asked the reader to examine this issue via a direct analysis of the second variation in part b) of Exercise 3.8. Since then, however, we learned a general method for establishing optimality—namely, the sufficient condition for optimality from Section 5.1.4—and it is instructive to see it in action here.

Specialized to our present LQR problem, the HJB equation (5.10) becomes

$$-V_t(t,x) = \inf_{u \in \mathbb{R}^m} \left\{ x^T Q(t)x + u^T R(t)u + \langle V_x(t,x), A(t)x + B(t)u \rangle \right\} \quad (6.16)$$

and the boundary condition (5.3) reads

$$V(t_1, x) = x^T M x. \quad (6.17)$$

Since $R(t) > 0$, it is easy to see that the infimum of the quadratic function of u in (6.16) is a minimum and to calculate (similarly to how we arrived at the formula (6.3) earlier) that the minimizing control is

$$u = -\frac{1}{2} R^{-1}(t) B^T(t) V_x(t,x). \quad (6.18)$$

We substitute this control into (6.16) and, after some term cancellations, bring the HJB equation to the following form:

$$
\begin{aligned}
-V_t(t,x) = x^T Q(t)x + (V_x(t,x))^T A(t)x \\
- \frac{1}{4}(V_x(t,x))^T B(t) R^{-1}(t) B^T(t) V_x(t,x).
\end{aligned}
\tag{6.19}
$$

In order to apply the sufficient condition for optimality proved in Section 5.1.4, we need to find a solution $V(\cdot,\cdot)$ of (6.19). Then, for the feedback law (6.12) to be optimal, it should match the feedback law given by (6.18) for this V. (The precise meaning of the last statement is provided by the formula (5.22) on page 165.) This will in turn be true if we have

$$
\frac{1}{2} V_x(t,x) = P(t)x
\tag{6.20}
$$

for all t and x. The equation (6.20) suggests that we should look for a function V that is defined in terms of P. Another observation that supports this idea is that in view of (6.11), the boundary condition (6.17) for the HJB equation can be written as

$$
V(t_1,x) = x^T P(t_1)x.
\tag{6.21}
$$

Armed with these facts, let us apply a certain amount of "wishful thinking" in solving the partial differential equation (6.19). Namely, let us try to *guess* a function V that satisfies the simple conditions (6.20) and (6.21), and then see if it satisfies the complicated equation (6.19). If it does, then it must be the value function and by the previous reasoning our control (6.12) must be optimal, hence everything will be proved. (In this last step, we are using the fact that (6.12) is a feedback law which does not depend on the initial condition; see the remarks immediately following the proof of the sufficient condition for optimality on page 167.) Our argument will be valid for all $t \le t_1$ for which $P(t)$ exists.

For the moment, let us proceed under the assumption (which will be validated momentarily) that $P(t)$ is symmetric for all t. A guess for V is actually fairly obvious, and might have already occurred to the reader:

$$
V(t,x) = x^T P(t)x.
\tag{6.22}
$$

This function clearly satisfies (6.21), and since its gradient with respect to x is $V_x(t,x) = 2P(t)x$ we see that (6.20) is also fulfilled. Now, let us check whether the function (6.22) satisfies (6.19). Since $V \in \mathcal{C}^1$, the viscosity solution concept is not needed here. Noting that $V_t(t,x) = x^T \dot{P}(t)x$ and plugging the two expressions for the partial derivatives of V into (6.19), we obtain

$$
-x^T \dot{P}(t)x = x^T Q(t)x + 2x^T P(t)A(t)x - x^T P(t)B(t)R^{-1}(t)B^T(t)P(t)x
$$

or, equivalently,

$$-x^T \dot{P}(t)x = x^T \big(Q(t) + P(t)A(t) + A^T(t)P(t) - P(t)B(t)R^{-1}(t)B^T(t)P(t)\big)x. \tag{6.23}$$

Since $P(t)$ satisfies the RDE (6.14), it immediately follows that (6.23) is a true identity. We conclude that, indeed, the function (6.22) is the value function (optimal cost-to-go) and the linear feedback law (6.12) is the optimal control. (We already know from Section 6.1.1 that an optimal control must be unique, and we also know that the sufficient condition of Section 5.1.4 guarantees global optimality.)

It is useful to reflect on how we found the optimal control. First, we singled out a candidate optimal control by using the maximum principle. Second, we identified a candidate value function and verified that this function and the candidate control satisfy the sufficient condition for optimality. Thus we followed the typical path outlined in Section 5.1.4. The next exercise takes a closer look at properties of $P(t)$ and closes a gap that we left in the above argument.

Exercise 6.2   *Let $t \le t_1$ be an arbitrary time at which the solution $P(t)$ of the RDE (6.14) with the boundary condition (6.11) exists.*

a) Prove that $P(t)$ is a symmetric matrix.

b) Prove that $P(t) \ge 0$ (positive semidefinite).

c) Can you prove that $P(t) > 0$ (positive definite)? If not, can you prove this by strengthening one of the standing assumptions?   □

It was both insightful and convenient to employ the sufficient condition for optimality in terms of the HJB equation to find the expression for the optimal cost and confirm optimality of the control (6.12). However, having the solution $P(t)$ of the RDE (6.14) with the boundary condition (6.11) in hand, it is also possible to solve the LQR problem by direct algebraic manipulations without relying on any prior theory.

Exercise 6.3   *Confirm the facts that (6.12) is the unique optimal control and (6.22) is the value function by using nothing more than square completion.*   □

6.1.4   Global existence of solution for the RDE

We are finally in a position to prove that the solution $P(\cdot)$ of the RDE (6.14), propagated backward in time from the terminal condition (6.11), exists for all $t \le t_1$. It follows from the standard theory of local existence and uniqueness of solutions to ordinary differential equations that $P(t)$ exists for t sufficiently close to t_1. (We already came across such results in Section 3.3.1, with the difference that there solutions were propagated forward in time; we also know about local existence of $P(t)$ from a different argument based on the formula (6.10), which we gave on page 183.) As for global existence, the problem is that some entry of P may have a finite escape time. In other words, there may exist a time $\bar{t} < t_1$ and some indices $i, j \in \{1, \ldots, n\}$ such that $P_{ij}(t)$ approaches $\pm\infty$ as $t \searrow \bar{t}$. Such behavior is actually quite typical for solutions of quadratic differential equations of Riccati type, as we discussed in detail on page 64. In the context of the formula (6.10), this would mean that the matrix $\Phi_{11}(t, t_1) - 2\Phi_{12}(t, t_1)M$ becomes singular at $t = \bar{t}$. Similarly, in the context of the formula (6.15) the finite escape would mean that the matrix $X(t)$ becomes singular at $t = \bar{t}$. In Section 2.6.2 we encountered a closely related situation and formalized it in terms of existence of conjugate points. Fortunately, in the present setting it is not very difficult to show by a direct argument that all entries of $P(t)$ remain bounded for all t, relying on the fact—established in the previous subsection—that $x^T P(t)x$ is the optimal LQR cost-to-go from (t, x) as long as it exists.

Seeking a contradiction, suppose that there is a $\bar{t} < t_1$ such that $P(t)$ exists on the interval $(\bar{t}, t_1]$ but some entry of $P(t)$ becomes unbounded as $t \searrow \bar{t}$. We know from Exercise 6.2 that for all $t \in (\bar{t}, t_1]$ the matrix $P(t)$ is symmetric and positive semidefinite, hence all its principal minors must be nonnegative. If an off-diagonal entry $P_{ij}(t)$ becomes unbounded as $t \searrow \bar{t}$ while all diagonal entries stay bounded, then a certain 2×2 principal minor of $P(t)$ must be negative for t sufficiently close to $\bar{t}$; namely, this is the determinant of the matrix

$$
\begin{array}{cc}
 & \begin{array}{cc} i & j \end{array} \\
\begin{array}{c} i \\ j \end{array} & \left(\begin{array}{cc} * & P_{ij}(t) \\ P_{ij}(t) & * \end{array} \right)
\end{array}
$$

formed by the i-th and the j-th row and column of $P(t)$. Thus this scenario is ruled out, and the only remaining possibility is that a diagonal entry $P_{ii}(t)$ becomes unbounded as $t \searrow \bar{t}$. Consider the vector $e_i := (0, \ldots, 1, \ldots, 0)^T \in \mathbb{R}^n$, with 1 in the i-th position and zeros everywhere else; then $e_i^T P(t)e_i = P_{ii}(t) \to \infty$ as $t \searrow \bar{t}$. Suppose that the system is in state e_i at some time $t > \bar{t}$. We know that $e_i^T P(t)e_i$ is the optimal cost-to-go from there, and so this cost must be unbounded as we take t to be closer to $\bar{t}$. On the other hand, this cost cannot exceed the cost of applying, e.g., the zero control on $[t, t_1]$. The state trajectory corresponding to this control is $x(s) = \Phi_A(s, t)e_i$ for $s \in [t, t_1]$, where $\Phi_A(\cdot, \cdot)$ is the state transition matrix for $A(\cdot)$, and the associated cost is

$$
\int_t^{t_1} \left(e_i^T \Phi_A^T(s, t)Q(s)\Phi_A(s, t)e_i \right) ds + e_i^T \Phi_A^T(t_1, t)M\Phi_A(t_1, t)e_i.
$$

It is quite clear that this cost remains bounded as t approaches $\bar{t}$, since the quantity inside the integral is bounded for $\bar{t} \leq t \leq s \leq t_1$. The resulting contradiction proves that a finite escape time $\bar{t}$ cannot exist.

The existence of the solution $P(\cdot)$ to the RDE (6.14) on the interval $[t_0, t_1]$ is now established. Thus we can be sure that the optimal control (6.12) is well defined, and the finite-horizon LQR problem has been completely solved. We must be able to explicitly solve the RDE, though, if we want to obtain a closed-form expression for the optimal control law.

Example 6.1   *To obtain the simplest possible finite-horizon LQR problem, consider the standard (scalar) integrator $\dot{x} = u$ and let the cost be $J(u) = \int_{t_0}^{t_1} (x^2(t) + u^2(t))dt$. All the matrices become scalars here: $A = 0$, $B = 1$, $Q = 1$, $R = 1$, $M = 0$. The RDE is the scalar differential equation $\dot{P} = P^2 - 1$, with the boundary condition $P(t_1) = 0$. Our theory tells us that its solution exists globally backward in time.[2] Let us see if we can compute this solution explicitly. Separating the variables and integrating, we have*

$$\int_{P(t)}^0 \frac{dP}{P^2 - 1} = \int_t^{t_1} ds.$$

The integral on the left-hand side evaluates to $\tanh^{-1} P(t)$, hence the solution we were seeking is $P(t) = \tanh(t_1 - t)$. Consequently, the feedback law $u(t) = -\tanh(t_1 - t)x(t)$ is the optimal control. $\square$

If analytically solving the RDE is not a completely trivial task even for such an elementary example, we expect closed-form solutions to be obtainable only in very special cases. Yet, the LQR problem is much more tractable compared to the general optimal control problem studied in Chapter 5. The main simplification is that instead of trying to solve the HJB equation which is a partial differential equation, we now have to solve the RDE which is an ordinary differential equation, and this can be done efficiently by standard numerical methods. In the next section, we define and study a variant of the LQR problem which lends itself to an even simpler solution; this development will be in line with what we already saw, in a more general context but in much less detail, towards the end of Section 5.1.3.

6.2   INFINITE-HORIZON LQR PROBLEM

The infinite-horizon version of the LQR problem is a special case of the general infinite-horizon problem constructed in Section 5.1.3. Starting with

---

[2]It is interesting to digress briefly and note that if we had $R = -1$ then the RDE would be $\dot{P} = -P^2 - 1$, which looks very similar but is fundamentally different in that its solutions do *not* exist globally backward in time (cf. page 64). Thus our standing assumption that R is positive is important. We invite the reader to figure out exactly where this assumption comes into play in the preceding argument.

the finite-horizon LQR problem defined in Section 6.1, we first assume that both the control system and the cost functional are time-invariant and that there is no terminal cost; this simply means that A, B, Q, and R are now constant matrices and $M = 0$. Then, we want to consider the limit as the final time t_1 approaches ∞. The resulting problem does not directly fit into the basic problem formulation of Section 3.3 and its well-posedness needs to be proved; in particular, we do not know a priori whether the optimal cost is finite (cf. footnote 2 on page 164). We did not try to settle this issue in the general context of Section 5.1.3, but we will do it here for the LQR problem. In this section we assume that the reader is familiar with basic concepts of linear system theory such as controllability and observability.

In preparation for studying the limit as $t_1 \to \infty$, let us treat the final time t_1 as a parameter which, unlike in Section 6.1, is no longer fixed. We want to make our notation more explicit by displaying the dependence of relevant quantities on this parameter. Specifically, from now on let us write V^{t_1} for the value function and denote by $P(t, t_1)$ the solution at time t of the RDE (6.14) with the boundary condition $P(t_1) = 0$. For each t_1, let us also relabel the optimal control and the optimal state trajectory (passing through the given initial condition x_0 at time t_0) for the corresponding finite-horizon LQR problem as $u^*_{t_1}$ and $x^*_{t_1}$, respectively. In this notation, the results of Section 6.1.3 say that

$$u^*_{t_1}(t) = -R^{-1}B^T P(t, t_1) x^*_{t_1}(t) \tag{6.24}$$

and

$$V^{t_1}(t, x) = x^T P(t, t_1) x. \tag{6.25}$$

In particular,

$$V^{t_1}(t_0, x_0) = x_0^T P(t_0, t_1) x_0 \tag{6.26}$$

is the finite-horizon optimal cost.

6.2.1   Existence and properties of the limit

We begin by making a series of observations about the behavior of $P(t_0, t_1)$ as a function of t_1, with the goal of establishing that $\lim_{t_1 \to \infty} P(t_0, t_1)$ exists (under the additional assumption of controllability) and has some interesting properties. This path will eventually lead us to a complete solution of the infinite-horizon LQR problem.

MONOTONICITY. It is not hard to see that *the finite-horizon optimal cost $x_0^T P(t_0, t_1) x_0$ is a monotonically nondecreasing function of the final time t_1.* Indeed, let $t_2 > t_1$. Using (6.25), the definition of the value function, and the standing assumptions that $Q \geq 0$ and $R > 0$, we have

$$x_0^T P(t_0, t_2) x_0 = V^{t_2}(t_0, x_0) = \int_{t_0}^{t_2} \left((x_{t_2}^*)^T(t) Q x_{t_2}^*(t) + (u_{t_2}^*)^T(t) R u_{t_2}^*(t) \right) dt$$

$$\geq \int_{t_0}^{t_1} \left((x_{t_2}^*)^T(t) Q x_{t_2}^*(t) + (u_{t_2}^*)^T(t) R u_{t_2}^*(t) \right) dt \geq V^{t_1}(t_0, x_0) = x_0^T P(t_0, t_1) x_0.$$

BOUNDEDNESS. It is not true in general that the optimal cost $x_0^T P(t_0, t_1) x_0$ remains bounded as $t_1 \to \infty$. For example, if the system is $\dot{x} = x$ (no control) then its solutions are growing exponentials and the infinite-horizon cost is clearly unbounded. However, we now show that *the finite-horizon optimal cost $x_0^T P(t_0, t_1) x_0$ remains bounded as $t_1 \to \infty$ assuming that (A, B) is a controllable pair.* Indeed, controllability guarantees the existence of a time $\bar{t}$ and a control $\bar{u}$ that steers the state from x_0 at time t_0 to 0 at time $\bar{t}$. After time $\bar{t}$, set $\bar{u}$ equal to 0. This control yields a state trajectory $\bar{x}$ satisfying $\bar{x}(t) = 0$ for all $t \geq \bar{t}$, and so for every $t_1 \geq \bar{t}$ we have

$$x_0^T P(t_0, t_1) x_0 = V^{t_1}(t_0, x_0) \leq J(\bar{u}) = \int_{t_0}^{\bar{t}} \left(\bar{x}^T(t) Q \bar{x}(t) + \bar{u}^T(t) R \bar{u}(t) \right) dt.$$

Since the above integral does not depend on t_1, it provides a uniform bound for the optimal cost—a single bound that is valid for all sufficiently large t_1, as desired. We leave the controllability assumption in force for the rest of this chapter (except for Exercise 6.5 on page 198 where its necessity will be re-examined).

EXISTENCE OF THE LIMIT. From the previous two claims it immediately follows that $x_0^T P(t_0, t_1) x_0$ has a limit as $t_1 \to \infty$. It turns out that more is true, namely, *the matrix $\lim_{t_1 \to \infty} P(t_0, t_1)$ is well defined.* To see why, let us consider some specific initial conditions x_0 (we can do this because all the facts established so far are valid for arbitrary x_0). First, let $x_0 = e_i$ with e_i as defined on page 188 for some $i \in \{1, \ldots, n\}$. Then $x_0^T P(t_0, t_1) x_0 = P_{ii}(t_0, t_1)$, implying that each diagonal entry of $P(t_0, t_1)$ has a limit as $t_1 \to \infty$. Next, let $x_0 = e_i + e_j$ for some $i \neq j$. Recalling that $P(t_0, t_1)$ is symmetric (Exercise 6.2), we have $x_0^T P(t_0, t_1) x_0 = P_{ii}(t_0, t_1) + 2 P_{ij}(t_0, t_1) + P_{jj}(t_0, t_1)$, from which we can deduce that the off-diagonal entries of $P(t_0, t_1)$ converge as well. We can think of $\lim_{t_1 \to \infty} P(t_0, t_1)$ as the solution of the RDE (6.14) that, starting from the zero matrix, has flown backward for infinite time and reached *steady state*; Figure 6.1 should help visualize this situation.

PROPERTIES OF THE LIMIT. Since the RDE (6.14) has now become a time-invariant differential equation, its solution $P(t_0, t_1)$ actually depends only on the difference $t_1 - t_0$. Thus it is clear that the steady-state solution $\lim_{t_1 \to \infty} P(t_0, t_1)$, whose existence we just established, does not depend on

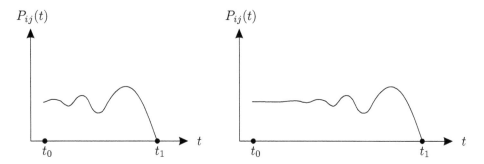

Figure 6.1: Steady-state solution of the RDE

t_0, i.e., it is a *constant matrix*. Denoting it simply by P, we have

$$P = \lim_{t_1 \to \infty} P(t, t_1) \qquad \forall\, t. \tag{6.27}$$

Next, passing to the limit as $t_1 \to \infty$ on both sides of the RDE (6.14), we see that $\lim_{t_1 \to \infty} \dot{P}(t, t_1)$ must also exist and be a constant matrix, which must then necessarily be the zero matrix. We thus conclude that P is a solution of the **algebraic Riccati equation (ARE)**

$$\boxed{PA + A^T P + Q - PBR^{-1}B^T P = 0} \tag{6.28}$$

Conceptually, the step of passing from the RDE (6.14), which is a matrix differential equation, to the ARE (6.28), which is a *static* matrix equation, mirrors our earlier step of passing from the general HJB equation (5.10) to its infinite-horizon counterpart (5.19). In both cases, the time derivative is eliminated; in the present case, we are left with no derivatives at all! The ARE (6.28) can be solved analytically or numerically without difficulties. It may happen, though, that the ARE has "spurious" extra solutions other than (6.27). In this regard, it is useful to note that the matrix P given by (6.27) must be *symmetric positive semidefinite* (because so is $P(t, t_1)$ for each t_1). We can hope that the ARE has only one solution with this additional property. This is the case in the next exercise, and we will show in Section 6.2.4 that this is always the case under appropriate assumptions.

Exercise 6.4 *Consider the double integrator $\dot{x}_1 = x_2$, $\dot{x}_2 = u$ and the cost $J(u) = \int_{t_0}^{\infty} (x_1^2(t) + x_2^2(t) + u^2(t))dt$. Find P by solving the ARE (6.28). Verify (either analytically or numerically) that it is indeed the steady-state solution of the RDE. Is the validity of the last statement affected by changing the terminal condition for the RDE (i.e., picking $M \neq 0$)?* □

6.2.2 Infinite-horizon problem and its solution

We have now prepared the ground for taking the limit as $t_1 \to \infty$ and considering the infinite-horizon LQR problem with the cost

$$J(u) = \int_{t_0}^{\infty} \left(x^T(t)Qx(t) + u^T(t)Ru(t) \right) dt. \tag{6.29}$$

Recalling the optimal cost (6.26) and the optimal control (6.24) for the finite-horizon case, and passing to the limit as $t_1 \to \infty$, it is natural to guess that the infinite-horizon optimal cost and optimal control will be[3]

$$V(x_0) = x_0^T P x_0 \tag{6.30}$$

and

$$u^*(t) = -R^{-1}B^T P x^*(t) \tag{6.31}$$

where P is the matrix limit (6.27) which satisfies the ARE (6.28). Note that the quadratic cost (6.30) is independent of t_0 and the linear feedback law (6.31) is time-invariant, which is consistent with the problem formulation and with our earlier findings in Section 5.1.3. Still, optimality of (6.31) is far from obvious. Strictly speaking, the use of the asterisks in (6.31) is not yet justified; at this point, x^* is simply the trajectory of the system under the action of the feedback law (6.31) with $x^*(t_0) = x_0$.

We now show that the above guess is indeed correct. Consider the function $\widehat{V}(x) := x^T P x$. Its derivative along the trajectory x^* is

$$\frac{d}{dt}\widehat{V}(x^*(t)) = (x^*)^T(t)\left(P(A - BR^{-1}B^T P) + (A^T - PBR^{-1}B^T)P\right)x^*(t)$$

$$= (x^*)^T(t)(PA + A^T P - 2PBR^{-1}B^T P)x^*(t)$$

$$= -(x^*)^T(t)(Q + PBR^{-1}B^T P)x^*(t)$$

where the last equality follows from the ARE (6.28). We can then calculate the portion of the corresponding cost over an arbitrary finite interval $[t_0, T]$ to be

$$\int_{t_0}^{T} \left((x^*)^T(t)Qx^*(t) + (u^*)^T(t)Ru^*(t) \right) dt$$

$$= \int_{t_0}^{T} (x^*)^T(t)(Q + PBR^{-1}B^T P)x^*(t)dt = -\int_{t_0}^{T} \frac{d}{dt}\widehat{V}(x^*(t))dt$$

$$= \widehat{V}(x_0) - \widehat{V}(x^*(T)) = x_0^T P x_0 - (x^*)^T(T)Px^*(T) \le x_0^T P x_0$$

---

[3]For consistency with our previous notation, it would have been more accurate to denote the limiting quantities by V^∞, u_∞^*, etc. We omit the superscripts and subscripts ∞ for simplicity, since we will be focusing on the infinite-horizon problem from now on.

where the last inequality follows from the fact that $P \geq 0$. Taking the limit as $T \to \infty$, we obtain

$$J(u^*) \leq x_0^T P x_0. \tag{6.32}$$

In particular, we can be sure that the infinite-horizon problem is well posed, because u^* gives a bounded cost. (The control $\bar{u}$ constructed on page 191 gives a bounded cost as well.) On the other hand, consider another trajectory x with the same initial condition corresponding to an arbitrary control u. Since $x_0^T P(t_0, t_1) x_0$ is the finite-horizon optimal cost, we have for every finite t_1 that

$$x_0^T P(t_0, t_1) x_0 \leq \int_{t_0}^{t_1} \left(x^T(t) Q x(t) + u^T(t) R u(t) \right) dt$$

$$\leq \int_{t_0}^{\infty} \left(x^T(t) Q x(t) + u^T(t) R u(t) \right) dt = J(u)$$

where the second inequality relies on the positive (semi)definiteness of Q and R. Passing to the limit as $t_1 \to \infty$ yields

$$x_0^T P x_0 \leq J(u).$$

Comparing this inequality with (6.32) and remembering that u was arbitrary (and could in particular be equal to u^*), we see that

$$J(u^*) = x_0^T P x_0 \leq J(u) \qquad \forall u$$

hence $x_0^T P x_0$ is the infinite-horizon optimal cost and u^* is an optimal control, as claimed.

6.2.3   Closed-loop stability

In the previous subsection we were able to obtain a complete solution to the infinite-horizon LQR problem, under the assumption (enforced since Section 6.2.1) that the system is controllable. Here we investigate an important property of the optimal closed-loop system which will motivate us to introduce one more assumption.

Example 6.2 *Consider the scalar system $\dot{x} = x + u$ and the cost $J(u) = \int_{t_0}^{\infty} u^2(t) dt$. The optimal control is, quite obviously, $u^* \equiv 0$ because it gives the zero cost. It corresponds to $P = 0$. (Incidentally, the ARE has another solution $P = 2$ which must be discarded.) The optimal closed-loop system $\dot{x}^* = x^*$ is unstable.* □

An optimal control that causes the state to grow unbounded is hardly acceptable. The reason for this undesirable situation in the above example is that the cost only takes into account the control effort and does not penalize instability. It is natural to ask under what additional assumption(s) the optimal control (6.31) automatically ensures that the state converges to 0. One option is to require Q to be strictly positive definite, but we will see in a moment that this would be an overkill. It is well known and easy to show (for example, via diagonalization) that every symmetric positive semidefinite matrix Q can be written as

$$Q = C^T C$$

where C is a $p \times n$ matrix with $p \le n$. Introducing the auxiliary output

$$y := Cx$$

we can rewrite the cost (6.29) as

$$J(u) = \int_{t_0}^{\infty} \left(y^T(t)y(t) + u^T(t)Ru(t) \right) dt.$$

Let us now assume that our system is observable through this output, i.e., *assume that (A, C) is an observable pair*. Note that if $Q > 0$ then $C = Q^{1/2}$ (matrix square root) which is $n \times n$ and nonsingular, and the observability assumption automatically holds. On the other hand, in Example 6.2 we had $Q = 0$ hence $C = 0$ and the observability assumption is violated.

It is not difficult to see why observability guarantees that the optimal closed-loop system is asymptotically stable. We know from (6.32) that the optimal control u^* gives a bounded cost (for arbitrary x_0). This cost is

$$J(u^*) = \int_{t_0}^{\infty} \left((y^*)^T(t)y^*(t) + (u^*)^T(t)Ru^*(t) \right) dt \qquad (6.33)$$

where $y^*(t) = Cx^*(t)$ is the output along the optimal trajectory. Next, recall that u^* is generated by the linear feedback law (6.31), which means that the optimal closed-loop system is a linear time-invariant system. Its solution x^* is thus given by a linear combination of (complex) exponential functions of time, and the same is true about y^* and u^*. This observation and the fact that $R > 0$ make it clear that in order for the integral in (6.33) to be bounded we must have $y^*(t) \to 0$ and $u^*(t) \to 0$ as $t \to \infty$. It is well known that if the output and the input of an observable linear time-invariant system converge to 0, then so does the state. (One way to see this is as follows: by observability, there exists a matrix L such that $A - LC$ has arbitrary desired eigenvalues with negative real parts; rewriting the system dynamics as $\dot{x} = (A - LC)x + Ly + Bu$, it is easy to show that $x \to 0$ when $y, u \to 0$.) Therefore, the optimal closed-loop system must be asymptotically—in fact, exponentially—stable.

6.2.4 Complete result and discussion

We now collect the results obtained in this section so far, as well as a few additional claims not yet proved, into a single theorem summarizing the solution of the infinite-horizon LQR problem and its main properties.

THEOREM 6.1 (Main results for the infinite-horizon LQR problem). *Consider the linear time-invariant control system*

$$\dot{x} = Ax + Bu$$

and the cost functional

$$J(u) = \int_{t_0}^{\infty} \left(x^T(t) C^T C x(t) + u^T(t) R u(t) \right) dt$$

where (A, B) is a controllable pair, (A, C) is an observable pair, and R is symmetric positive definite (all vectors and matrices have arbitrary compatible dimensions). Then the following statements are true:

1) *The limit $P := \lim_{t_1 \to \infty} P(t_0, t_1)$ of the solution at time t_0 of the RDE (6.14) with the boundary condition $P(t_1) = 0$ exists; this is a constant matrix (independent of t_0) and it is a unique symmetric positive definite solution of the ARE (6.28).*

2) *The optimal cost is $V(x_0) = x_0^T P x_0$, which is the limit as $t_1 \to \infty$ of the finite-horizon optimal cost (6.26).*

3) *The unique optimal control has the linear time-invariant state feedback form $u^*(t) = -R^{-1} B^T P x^*(t)$, which is the limit as $t_1 \to \infty$ of the finite-horizon optimal feedback law (6.24).*

4) *The closed-loop system $\dot{x}^* = (A - B R^{-1} B^T P) x^*$ is exponentially stable.*

PROOF. Statement 1 of the theorem was proved in Section 6.2.1 with the exception of two claims: the strict positive definiteness of P (we only showed that $P = P^T \geq 0$) and the uniqueness property. Statements 2 and 3 were proved in Section 6.2.2, except for the uniqueness of the optimal control. Statement 4 was proved in Section 6.2.3. With these results in place, we now establish the remaining claims.

Let us first prove that $P > 0$. We already know that $P \geq 0$. Suppose that $x_0^T P x_0 = 0$ for some x_0, which in view of statement 2 means that for this initial condition x_0 the optimal cost $\int_{t_0}^{\infty} ((x^*)^T(t) C^T C x^*(t) + (u^*)^T(t) R u^*(t)) dt$ equals 0. This is possible only if $Cx^* \equiv 0$ and $u^* \equiv 0$ (since $R > 0$). By observability we must then have $x^* \equiv 0$, as it is well known that the output of an observable linear time-invariant system (with the zero input) can be identically 0 only along the zero trajectory. Thus $x_0 = 0$ which proves that P is indeed positive definite.

Let us now prove that P is a unique solution of the ARE in the class of positive definite matrices. In fact, we will show that it is unique even in the class of positive semidefinite matrices. Suppose that the ARE has another positive semidefinite solution $\bar{P}$. Consider the new cost

$$\bar{J}^{t_1}(u) := \int_{t_0}^{t_1} \left(x^T(t)Qx(t) + u^T(t)R(t)u(t) \right)dt + x^T(t_1)\bar{P}x(t_1)$$

and its infinite-horizon counterpart

$$\bar{J}^\infty(u) := \lim_{t_1 \to \infty} \left(\int_{t_0}^{t_1} \left(x^T(t)Qx(t) + u^T(t)R(t)u(t) \right)dt + x^T(t_1)\bar{P}x(t_1) \right).$$

By statements 4 and 2 of this theorem, the same control u^* as in statement 3 gives

$$\bar{J}^\infty(u^*) = \int_{t_0}^\infty \left((x^*)^T(t)Qx^*(t) + (u^*)^T(t)Ru^*(t) \right)dt = x_0^T P x_0$$

and this is the optimal cost with respect to $\bar{J}^\infty$ because for every other control u we have

$$\bar{J}^\infty(u) \geq \int_{t_0}^\infty \left(x^T(t)Qx(t) + u^T(t)R(t)u(t) \right)dt \geq x_0^T P x_0.$$

We also know from the results of Section 6.1 that the optimal cost with respect to $\bar{J}^{t_1}(u)$ is $x_0^T P(t_0; \bar{P}, t_1)x_0$, where $P(t_0; \bar{P}, t_1)$ denotes the solution at time t_0 of the RDE (6.14) with the boundary condition $P(t_1) = \bar{P}$. But this solution must be $\bar{P}$ itself, for every t_1, because $\bar{P}$ is an equilibrium solution of the RDE (by virtue of satisfying the ARE). It follows[4] that $\bar{P} = P$, and the uniqueness property is confirmed.

Finally, to show that the optimal control is unique, recall the equations (5.13) and (5.14) in Section 5.1.3 which say that an optimal control must satisfy

$$u^*(t) = \arg\min_{u \in U} \left\{ L(t, x^*(t), u) + \langle V_x(t, x^*(t)), f(t, x^*(t), u) \rangle \right\}$$

or, what is the same,

$$u^*(t) = \arg\max_{u \in U} H(t, x^*(t), u, -V_x(t, x^*(t))).$$

---

[4]This step is actually not completely obvious; we leave the details for the reader to work out.

For the infinite-horizon LQR problem and the value function $V(x) = x^T P x$ this condition takes the form

$$u^*(t) = \arg \min_{u \in \mathbb{R}^m} \left\{ (x^*)^T(t) Q x^*(t) + u^T R u \right.$$
$$\left. + 2(x^*)^T(t) P A x^*(t) + 2(x^*)^T(t) P B u \right\}$$

and it is easy to check (cf. Section 6.1.3) that this uniquely identifies the feedback law given in statement 3. □

We can see from the proof that the observability assumption was only used for establishing exponential stability of the optimal closed-loop system (statement 4) and the positive definiteness and uniqueness of P in statement 1 (the uniqueness proof relied on observability indirectly because it invoked statement 4). The controllability assumption, on the other hand, was used for showing the existence of P which is crucial for all the other claims. Nevertheless, there remains the possibility that some or all claims could be proved without relying on these assumptions.

Exercise 6.5   *Suppose that the assumptions of controllability of (A, B) and observability of (A, C) are replaced by stabilizability and detectability, respectively. Examine the validity of each claim of Theorem 6.1. (If still true, justify; otherwise, explain why not and how the statement should be modified to become true.)* □

In view of statement 1 of Theorem 6.1 and its proof, we know that the ARE has exactly one positive semidefinite solution P, and this solution is in fact positive definite and gives the optimal cost and optimal control via statements 2 and 3. For example, consider the infinite-horizon version of Example 6.1, with the system $\dot{x} = u$ and the cost $J(u) = \int_{t_0}^{\infty} (x^2(t) + u^2(t)) dt$. The ARE $P^2 - 1 = 0$ has two solutions, $P = \pm 1$, of which $P = 1$ is the one we want. The optimal cost is $V(x_0) = x_0^2$ and the optimal control is $u^* = -x^*$. The result that the reader obtained in Exercise 6.4 should also immediately yield the optimal cost for that problem and the exponentially stable optimal closed-loop system.

Linearity, time-invariance, and exponential stability are very desirable features of the optimal closed-loop system, indicating that the infinite-horizon LQR problem provides good control design guidelines. The choice of the matrices Q and R is often part of the design process, which helps shape the behavior of the state and the control signal.

Exercise 6.6   *Consider the scalar system $\dot{x} = ax + bu$ and the cost $J(u) = \int_{t_0}^{\infty} (qx^2(t) + ru^2(t)) dt$, where $a, q, r > 0$ and b is arbitrary. Suppose that a, b, q are fixed but r can vary. Show that for $r \to 0$ (the "cheap control" case) the eigenvalue of the optimal closed-loop system moves off to $-\infty$, while for*

$r \to \infty$ (the "expensive control" case) the eigenvalue of the optimal closed-loop system tends to $-a$, i.e., the opposite of the open-loop eigenvalue. $\square$

6.3   NOTES AND REFERENCES FOR CHAPTER 6

The linear quadratic regulator problem is a classical subject covered in many textbooks. This chapter has particularly benefited from the expositions in [AF66, AM90, Bro70, KS72]; basic background facts from linear system theory can also be found in these books, as well as in more recent texts such as [Hes09]. Kalman's original LQR paper [Kal60] is a must-read, as it contains the first treatment of controllability for linear systems and a few other fundamental concepts and techniques. Among several possible ways of deriving the solution to the finite-horizon LQR problem, we favor the approach followed in [AF66] and [KS72] where the linear state feedback form of the optimal control is established independently of the Riccati differential equation (which is derived later). Section 24 of [Bro70] contains an insightful discussion of Riccati differential equations and their solutions. Various methods for numerical solution of the RDE are described in [AM90, Appendix E], [KS72, Section 3.5], and the references therein. For the infinite-horizon LQR problem, different sources again take different routes to arrive at the main results. In particular, [AM90] gives a purely linear-algebraic proof of closed-loop stability which does not rely on the Lyapunov argument that we used in Section 6.2.2.

Many generalizations of the LQR problem are not discussed here. These include: more general quadratic costs involving cross-terms; tracking problems; the output feedback version of the LQR problem and its relation to optimal state estimation (the Kalman filter). The references [AF66, AM90, KS72] cited earlier can be consulted for detailed information on these topics. In Section 7.3 we will discuss robust control problems which can be considered as extensions of the LQR problem.

Chapter Seven

Advanced Topics

In this chapter we discuss four additional topics (independent from one another) which extend the developments of Chapters 4–6. The objective of the present chapter is twofold: to enhance the reader's understanding of the earlier material, and to indicate some of the possible avenues for further study. We will not strive for the same level of completeness and rigor here as in the previous chapters, and instead will limit ourselves to an introductory discussion. The references provided at the end of the chapter, as usual, can be consulted for precise statements of technical results and other details.

7.1 MAXIMUM PRINCIPLE ON MANIFOLDS

Ever since we formulated the general optimal control problem in Section 3.3, we have allowed the state x to take arbitrary values in $\mathbb{R}^n$. However, in many situations of interest $\mathbb{R}^n$ is not an adequate state space (just like $U = \mathbb{R}^m$ is not always an adequate control space). Recall, for example, the pendulum dynamics (2.55) that we derived on page 57. The angle variable θ naturally lives on a circle. Combining it with the angular velocity $\dot{\theta} \in \mathbb{R}$, we obtain the state that evolves on a cylinder. State spaces of this kind are quite typical for mechanical systems.

Surfaces such as a sphere, a cylinder, or a torus are all examples of (differentiable) *manifolds*. The purpose of this section is to reformulate our optimal control problem and the maximum principle in the geometric language of manifolds. Besides a higher level of generality, casting the maximum principle in the framework of manifolds has another important benefit: it clarifies the intrinsic meaning of the costate (adjoint vector), thereby greatly elucidating the essence of the maximum principle even in the familiar setting of control problems in $\mathbb{R}^n$. We begin this task in Section 7.1.1, where we describe manifolds more precisely as surfaces defined by equality constraints and discuss some fundamental objects associated with manifolds. It is worth noting that scenarios in which a state space is characterized by inequality constraints, although equally significant, are not captured by the present set-up.

7.1.1   Differentiable manifolds

We say that a set M is a k-dimensional differentiable manifold embedded in $\mathbb{R}^n$, or simply a *manifold*, if it is given by

$$M = \{x \in \mathbb{R}^n : h_1(x) = h_2(x) = \cdots = h_{n-k}(x) = 0\} \qquad (7.1)$$

where h_i, $i = 1, \ldots, n-k$ are $\mathcal{C}^1$ functions from $\mathbb{R}^n$ to $\mathbb{R}$ such that the gradient vectors $\nabla h_i(x)$, $i = 1, \ldots, n-k$ are linearly independent for each $x \in M$. This is actually not a new object for us: it is just a k-dimensional surface in $\mathbb{R}^n$ already considered in Section 1.2.2 and later in the formulation of the Basic Variable-Endpoint Control Problem in Chapter 4. The linear independence assumption imposed on the gradients means, in the terminology of Section 1.2.2, that all points of M are regular.

Intuitively speaking, a k-dimensional manifold M is a subset of $\mathbb{R}^n$ that locally "looks like" $\mathbb{R}^k$. This idea can be made precise by equipping M with *local coordinates*, as follows. Fix an arbitrary $x \in M$. Since by assumption the (nonsquare) Jacobian matrix

$$\begin{pmatrix} (h_1)_{x_1}(x) & \cdots & (h_1)_{x_n}(x) \\ \vdots & \ddots & \vdots \\ (h_{n-k})_{x_1}(x) & \cdots & (h_{n-k})_{x_n}(x) \end{pmatrix}$$

is of rank $n - k$, it has $n - k$ linearly independent columns; shuffling the variables $x_1, \ldots, x_n$ if necessary, we can assume that these are the last $n - k$ columns. Using the Implicit Function Theorem, we can then solve the equations in (7.1) for the variables $x_{k+1}, \ldots, x_n$ in terms of $x_1, \ldots, x_k$ in some neighborhood of x. Consequently, the components $x_1, \ldots, x_k$ can be used to describe points in M near x, i.e., they provide a local coordinate chart for M. A simple example of a 1-dimensional manifold in $\mathbb{R}^2$ is the unit circle, for which we invite the reader to work out the above construction.

The standard definition of a manifold is actually stated in terms of the existence of local coordinate charts satisfying suitable compatibility conditions, without any explicit reference to the ambient space $\mathbb{R}^n$. However, it is known that every manifold defined in this way can be embedded in $\mathbb{R}^n$ for some n large enough. The above more concrete notion of an embedded manifold is sufficient for our purposes, but the concept of local coordinates will be useful for us as well. Basically, we have the choice of representing points in M as n-vectors or locally as k-vectors (cf. the discussion of holonomic constraints in Section 2.5.2).

Let M be a k-dimensional manifold. We know that at each point $x \in M$ we can define the *tangent space* $T_x M$ (see page 11). This is a k-dimensional linear vector space, spanned by the tangent vectors associated with all possible curves in M passing through x. Recall that for a curve $x(\cdot)$ lying in

M, with $x(0) = x$, the corresponding tangent vector is $\dot{x}(0)$. In local coordinates, if the curve has components $x_1(\cdot), \ldots, x_k(\cdot)$ then the components of the tangent vector $\xi = \dot{x}(0)$ are $(\xi_1, \ldots, \xi_k) = (\dot{x}_1(0), \ldots, \dot{x}_k(0))$. It is also useful to consider the union

$$TM := \bigcup_{x \in M} T_x M$$

which is called the *tangent bundle* of M. This is a manifold of dimension $2k$. We note that, in general, TM is not globally diffeomorphic to the direct product $M \times \mathbb{R}^k$ (well-known counterexamples are obtained by taking M to be the two-dimensional sphere or the Mobius band). It is convenient to think of vector fields on M as *sections* of the tangent bundle, i.e., functions $f : M \to TM$ such that $f(x) \in T_x M$ for all x.

Given a point x on a manifold M, the linear vector space of linear $\mathbb{R}$-valued functions on $T_x M$ is called the *cotangent space* to M at x, and is denoted by $T_x^* M$. This is the *dual space*[1] to $T_x M$; its elements are called *cotangent vectors*, or simply *covectors*. The simplest example of a covector is the differential of a function. Let $g : M \to \mathbb{R}$ be a $\mathcal{C}^1$ function. For each $x \in M$, the linear function $dg|_x : T_x M \to \mathbb{R}$ is defined as follows. Given a tangent vector $\xi \in T_x M$ and an arbitrary curve $x(\cdot)$ in M with $x(0) = x$ having ξ as its tangent vector (so that $\dot{x}(0) = \xi$), let

$$dg|_x (\xi) := \left. \frac{d}{d\alpha} \right|_{\alpha=0} g(x(\alpha)).$$

In local coordinates this is represented by $dg|_x (\xi) = \sum_{i=1}^{k} g_{x_i}(x) \dot{x}_i(0) = \sum_{i=1}^{k} g_{x_i}(x) \xi_i$, which clearly depends only on ξ and not on the choice of a specific curve $x(\cdot)$. The map $dg|_x$ assigns to each tangent vector ξ the derivative of g in the direction of ξ. If $x_1, \ldots, x_k$ are local coordinates on M, then the corresponding differentials $dx_1, \ldots, dx_k$ form a basis for $T_x^* M$ and hence yield local coordinates on the cotangent space. For a tangent vector $\xi \in T_x M$, the numbers $dx_i(\xi)$ give the components ξ_i of ξ. One also defines the *cotangent bundle*

$$T^* M = \bigcup_{x \in M} T_x^* M$$

which is itself a manifold, of dimension $2k$. Combining local coordinates $x_1, \ldots, x_k$ on M with local coordinates on $T_x^* M$ relative to the basis given by $dx_1, \ldots, dx_k$, one obtains local coordinates on the cotangent bundle $T^* M$ known as *canonical coordinates*.

---

[1]The use of an asterisk for the dual space is not to be confused with the use of the same symbol throughout the book to indicate optimality.

7.1.2   Re-interpreting the maximum principle

Suppose that we are given a (time-invariant) control system

$$\dot{x} = f(x, u) \tag{7.2}$$

whose state x takes values in some k-dimensional manifold M and whose control u takes values in some control set U. For its solution $x(\cdot)$ to stay in M, the velocity vector $f(x, u)$ must be tangent to M at x for all x and u. Thus we see that there is an important difference between the states and the velocity vectors: the former live in M while the latter live in the tangent bundle TM. When we worked over $\mathbb{R}^n$, which is its own tangent space, we never explicitly made this distinction. In local coordinates, the system description takes the form

$$\begin{pmatrix} \dot{x}_1 \\ \vdots \\ \dot{x}_k \end{pmatrix} = \begin{pmatrix} f_1(x, u) \\ \vdots \\ f_k(x, u) \end{pmatrix} \in \mathbb{R}^k$$

and the difference between states and tangent vectors becomes "hidden" once again.

Let us assume for simplicity that an optimal control problem is formulated in the Mayer form (i.e., with terminal cost only). We know that problems with running cost can always be converted to this form by appending an additional state x^0, which would yield a system on the augmented manifold $\mathbb{R} \times M$ (cf. Sections 3.3.2 and 4.2.1). The basic ingredients of the maximum principle are the costate p and the Hamiltonian H. In the case when $M = \mathbb{R}^n$, the Hamiltonian for the Mayer problem took the form $H(x, u, p) = \langle p, f(x, u) \rangle$. For a general manifold M, we need to ask ourselves which space p should belong to and how H should be (re)defined. Our first natural guess might be that p, like $f(x, u)$, should be a tangent vector to M. However, in contrast with $f(x, u)$, there is no clear geometric reason why p should be a tangent vector. Also, taking p to be a tangent vector, we cannot assign a new meaning to our earlier definition of H unless we equip the tangent space with an inner product. (Introducing an inner product on each tangent space $T_x M$—called a Riemannian metric on M—is possible but, as we will see, is neither necessary nor relevant for our present purposes.) Another option that might come to mind is that p should live in M itself; however, this choice offers even fewer clues towards any natural interpretation of the Hamiltonian.

Can we perhaps take a more direct guidance from the fact that in our old definition of the Hamiltonian, p appears in an inner product with the velocity vector $f(x, u)$? In fact, we already remarked in Section 3.4.2 that p never appears by itself but always inside inner products such as $\langle p, \dot{x} \rangle$; in other words, it *acts* on velocity vectors. This observation suggests that the intrinsic role of the costate p is not that of a tangent vector, but that of a *covector*. To better understand the difference between these two types of objects and why the latter one correctly captures the notion of a costate, let us look at how they propagate along a flow induced by a dynamical system on M.

Fix a number $\tau > 0$ and let $\Phi_\tau : M \to M$ be a C^1 map. While the construction that we are about to describe is valid for every such map, the map that we have in mind here is the one obtained by flowing forward for τ units of time along the trajectory of the system (7.2) corresponding to some fixed control u (which, ultimately, is taken to be an optimal control for a given initial condition). Let us first discuss the transformation that Φ_τ induces on tangent vectors. Pick a point $x \in M$ and a tangent vector $\xi \in T_x M$. We know that ξ is tangent to some curve in M passing through x, namely, $\xi = \dot{x}(0)$ where $x(\alpha) \in M$ for real α (around 0) and $x(0) = x$. The image $\Phi_\tau(x(\cdot))$ of this curve under the map Φ_τ is a curve in M which passes through $\Phi_\tau(x)$, as illustrated in Figure 7.1. Denote the tangent vector at $\Phi_\tau(x)$ associated with this new curve by $d\Phi_\tau|_x(\xi)$; in other words, define

$$d\Phi_\tau|_x(\xi) := \frac{d}{d\alpha}\bigg|_{\alpha=0} \Phi_\tau(x(\alpha)).$$

The above quantity depends only on the vector ξ and not on the choice of a particular curve $x(\cdot)$ with this tangent vector. In this way we obtain a natural definition of a linear map

$$d\Phi_\tau|_x : T_x M \to T_{\Phi_\tau(x)} M$$

called the *derivative* (or *differential*) of Φ_τ at x. It extends to manifolds the standard notion of the differential of a function (described by its Jacobian matrix) from vector calculus; more generally, functions from M to another manifold N can be considered, and we already discussed the case $N = \mathbb{R}$ in Section 7.1.1. In fact, in the proof of the maximum principle we performed a closely related computation, showing that an infinitesimal state perturbation ξ propagates along an optimal trajectory of (7.2) according to the variational equation

$$\dot{\xi} = f_x|_* \xi \tag{7.3}$$

(see Section 4.2.4). The derivative map $d\Phi_\tau$ pushes the tangent vectors forward in the direction of action of the original map Φ_τ on M. Objects such as tangent vectors, which propagate forward along a flow on M in this sense, are called *contravariant*.

Now suppose that we are given a *covector* at x, i.e., a linear function on $T_x M$. Let us denote it by $p|_x$ so as to have $p|_x(\xi) \in \mathbb{R}$ for each $\xi \in T_x M$. For the same map $\Phi_\tau : M \to M$ as before, can we define in a natural

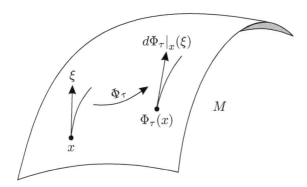

Figure 7.1: Tangent vectors propagate forward

way a linear function $p|_{\Phi_\tau(x)}$ on $T_{\Phi_\tau(x)}M$? We must decide what the value $p|_{\Phi_\tau(x)}(\eta)$ should be for every $\eta \in T_{\Phi_\tau(x)}M$. While it is tempting to say that $p|_{\Phi_\tau(x)}(\eta)$ should equal the value of $p|_x$ on the preimage of η under the map Φ_τ, this preimage is not well defined unless the map Φ_τ is invertible. In fact, the reader will quickly realize that there is no apparent candidate map for propagating covectors along Φ_τ similarly to how the derivative map $d\Phi_\tau$ acts on tangent vectors. The reason is that, instead of trying to push covectors forward, we should *pull them back*. This revised objective is readily accomplished as follows: given a covector $p|_{\Phi_\tau(x)}$ on $T_{\Phi_\tau(x)}M$, define a covector $p|_x$ on T_xM by

$$p|_x(\xi) := p|_{\Phi_\tau(x)}(d\Phi_\tau|_x(\xi)). \tag{7.4}$$

As we indicated earlier, the intended meaning of Φ_τ is that of flowing forward for τ units of time along an optimal trajectory of (7.2), and the infinitesimal (as $\tau \to 0$) transformation induced by the derivative map $d\Phi_\tau$ is the variational equation (7.3). We can now recognize the formula (7.4) as expressing—in an intrinsic, coordinate-free fashion—the adjoint property from Section 4.2.8; indeed, it guarantees that $p(\xi)$ stays constant along the trajectory. The familiar adjoint equation $\dot{p} = -(f_x)^T|_* p$ is nothing but the infinitesimal version of (7.4) written in local coordinates. A fact not really revealed by this differential equation is that covectors are *covariant* objects, in the sense that they propagate backward along a flow on M. This is exactly why we always have *terminal* rather than initial conditions for the costate!

Now everything is beginning to fall into place. The Hamiltonian for our Mayer problem on a manifold M should be defined as

$$H(x, u, p) = p(f(x, u)) \tag{7.5}$$

where the costate p is a covector at x (strictly speaking, it would be more accurate to write it as $p|_x$). The maximum principle postulates the existence of a costate $p^*(t) \in T^*_{x^*(t)}M$ for each t, where x^* is the optimal trajectory being analyzed. The terminal value $p^*(t_f)$ uniquely specifies $p^*(t)$ for all $t \in [t_0, t_f]$ as explained above. The Hamiltonian maximization condition takes the same form as in Chapter 4. For a formal statement of the maximum principle on manifolds along these lines, see the references listed in Section 7.5. There is, however, one more concept that is usually involved when such results are stated in the literature, and we examine it briefly in the next subsection.

7.1.3   Symplectic geometry and Hamiltonian flows

In our discussion of Hamiltonian mechanics in Section 2.4, the Hamiltonian was given a clear physical interpretation as the total energy of the system. Hamilton's canonical differential equations, on the other hand, were derived formally from the Euler-Lagrange equation and we never paused to consider their intrinsic meaning. We fill this gap here by exposing the important connection between symplectic geometry and Hamiltonian flows, which provides one further insight into the geometric formulation of the maximum principle.

Let us return to the cotangent bundle T^*M which, as explained at the end of Section 7.1.1, is a $2k$-dimensional manifold equipped with canonical local coordinates. On this manifold there is a natural differential 2-form ω^2, called a *symplectic form* (or *symplectic structure*). This is a bilinear skew-symmetric form on the tangent space to T^*M at each point; in other words, it acts on pairs of tangent vectors to T^*M (and, moreover, it depends smoothly on the choice of a point in T^*M). In canonical coordinates $(x, p) = (x_1, \ldots, x_k, p_1, \ldots, p_k)$ on T^*M, the symplectic form that we consider here is given by

$$\omega^2 := dx_1 \wedge dp_1 + \cdots + dx_k \wedge dp_k$$

where the exterior multiplication $\wedge$ is defined by

$$(\omega_1 \wedge \omega_2)(\xi_1, \xi_2) := \omega_1(\xi_1)\omega_2(\xi_2) - \omega_2(\xi_1)\omega_1(\xi_2).$$

(The exterior product is the oriented area spanned by the vectors $\begin{pmatrix} \omega_1(\xi_1) \\ \omega_1(\xi_2) \end{pmatrix}$ and $\begin{pmatrix} \omega_2(\xi_1) \\ \omega_2(\xi_2) \end{pmatrix}$, or the determinant of the corresponding matrix.) Given two tangent vectors to T^*M with components $(y, q) = (y_1, \ldots, y_k, q_1, \ldots, q_k)$ and $(z, r) = (z_1, \ldots, z_k, r_1, \ldots, r_k)$, it is straightforward to check that

$$\omega^2\big((y, q), (z, r)\big) = \sum_{i=1}^{k} (y_i r_i - q_i z_i). \tag{7.6}$$

The symplectic form allows us to establish a one-to-one correspondence between tangent vectors to T^*M and covectors on T^*M, as follows. To a tangent vector ξ (at a fixed point of T^*M) we associate a covector ω_ξ^1 which acts on another tangent vector η to T^*M according to

$$\omega_\xi^1(\eta) := \omega^2(\xi, \eta). \tag{7.7}$$

The map $\xi \mapsto \omega_\xi^1$ is a linear map between vector spaces, whose matrix with respect to the canonical basis is $\begin{pmatrix} 0 & -I \\ I & 0 \end{pmatrix}$.

Now, we can view our Hamiltonian (7.5) as a function on the cotangent bundle T^*M (i.e., a function of the variables x and p) parameterized additionally by the controls u. Then the differential dH of the Hamiltonian gives us a covector on T^*M at each point. Applying the inverse of the map constructed above to this covector, we obtain a tangent vector at each point of T^*M, or a vector field on T^*M. This vector field is called the *Hamiltonian vector field* and is denoted by $\vec{H}$. In canonical coordinates, the flow along $\vec{H}$ is described by the familiar differential equations

$$\dot{x} = H_p, \qquad \dot{p} = -H_x. \tag{7.8}$$

Indeed, for an arbitrary tangent vector η to T^*M with components $(z, r) = (z_1, \ldots, z_k, r_1, \ldots, r_k)$, we can easily check using (7.6) that with $\vec{H}$ defined via (7.8) we have

$$\omega^2(\vec{H}, \eta) = \sum_{i=1}^{k} (H_{x_i} z_i + H_{p_i} r_i) = dH(\eta)$$

and the desired conclusion follows from (7.7). Therefore, the statement of the maximum principle about the existence of a costate satisfying the adjoint equation (the second canonical equation) can be reformulated as saying that an optimal state trajectory can be *lifted* to an integral curve of the Hamiltonian vector field. (The control u remains an argument on the right-hand side of (7.8), but the maximum principle is of course applied with an optimal control u^* plugged in.)

7.2   HJB EQUATION, CANONICAL EQUATIONS, AND CHARACTERISTICS

We know from Section 5.1.3 that the HJB equation is a PDE which is difficult to solve in general. In Section 5.2 we compared the HJB equation with the necessary conditions of the maximum principle and demonstrated, in particular, how solving the HJB equation for the value function V formally leads to a solution of the canonical equations (via $p := -V_x$). In this section we discuss an important piece of the PDE theory which enables one, in principle, to solve a PDE with the help of a suitable system of ODEs called the *characteristics*. We will show that the canonical equations arise as the characteristics of the HJB equation. Thus, compared to Section 5.2, the method of characteristics takes us in the opposite direction: from a solution of the canonical equations to a solution of the HJB equation. We will first introduce this method in the context of general PDEs and then specialize it to the setting of the HJB equation.

7.2.1   Method of characteristics

Suppose that we are looking for a function $v : \mathbb{R}^n \to \mathbb{R}$ that solves the PDE

$$F(x, v(x), \nabla v(x)) = 0 \tag{7.9}$$

where $F : \mathbb{R}^n \times \mathbb{R} \times \mathbb{R}^n \to \mathbb{R}$ is a given continuous function. This is the same equation as (5.34) but here we assume that v is differentiable in the classical sense. We are especially interested in the situation where F depends linearly on ∇v; such PDEs are called *quasi-linear*. In addition, for simplicity and better geometric intuition, we start with the case when $n = 2$. Then (7.9) takes the form

$$a(x_1, x_2, v)v_{x_1} + b(x_1, x_2, v)v_{x_2} = c(x_1, x_2, v) \tag{7.10}$$

for some functions a, b, and c. Here and below, we omit some or all arguments of functions wherever convenient.

The graph of a solution $v = v(x_1, x_2)$ to (7.10) is a surface in the (x_1, x_2, v)-space defined by the equation $h(x_1, x_2, v) := v(x_1, x_2) - v = 0$. The gradient vector $\nabla h = (v_{x_1}, v_{x_2}, -1)^T$ is normal to this surface at each point. Noting that the PDE (7.10) can be equivalently written as

$$\left\langle \begin{pmatrix} a \\ b \\ c \end{pmatrix}, \begin{pmatrix} v_{x_1} \\ v_{x_2} \\ -1 \end{pmatrix} \right\rangle = 0$$

we conclude that the vector $(a, b, c)^T$ is everywhere orthogonal to the normal ∇h, hence it lies in the tangent space to the solution surface. This is the geometric interpretation of the quasi-linear PDE (7.10): the vector field $(a, b, c)^T$ is tangent to the solution surface $v = v(x_1, x_2)$ at each point. The system of ODEs associated with this vector field can be written as

$$\frac{dx_1}{ds} = a(x_1, x_2, v), \qquad \frac{dx_2}{ds} = b(x_1, x_2, v), \qquad \frac{dv}{ds} = c(x_1, x_2, v). \tag{7.11}$$

These are called the *characteristic ODEs* of the PDE (7.10), and their solution curves are called the *characteristics*. With a slight abuse of terminology, we sometimes use the term "characteristics" not only for the solution curves of the equations (7.11) but also for these equations themselves.

The characteristic curves "fill" our solution surface $v = v(x_1, x_2)$; skipping ahead to Figure 7.2 may help the reader visualize this situation. As an example, it is useful to think about the case when $c \equiv 0$. Then (7.10) says that the derivative of v in the direction of the vector $(a, b)^T$ is 0, implying that v stays constant along solutions of the equations $\dot{x}_1 = a$, $\dot{x}_2 = b$. Every function v whose level sets are solution curves of these two equations in the (x_1, x_2)-plane is a solution of the PDE, and the characteristics are horizontal slices of the corresponding solution surface. Of course, in general (for nonzero c) characteristics are not horizontal.

Now, in principle we should be able to solve the characteristic ODEs and obtain the characteristics. How does this help us solve the original PDE (7.10)? Without additional specifications, there are too many characteristics to identify a desired solution surface $v = v(x_1, x_2)$; indeed, there is a characteristic passing through every point in the (x_1, x_2, v)-space, so they fill the entire space. We know that for ODEs, to fix a particular solution we must specify a point through which it passes (an initial condition). The counterpart of this for the PDE (7.10) is that we must specify a suitable *curve* in the (x_1, x_2, v)-space through which the solution surface passes; we then obtain what is called a *Cauchy (initial value) problem*. Geometrically, the idea is that characteristics that pass through points on this initial curve will determine the solution surface of the PDE, as illustrated in Figure 7.2 (with the thick curve representing an initial curve).

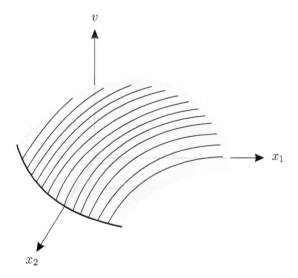

Figure 7.2: Characteristics for an initial value problem

An initial curve can be defined in parametric form as

$$x_1 = x_1(r), \qquad x_2 = x_2(r), \qquad v = v(r)$$

where $r \in \mathbb{R}$. Then, the desired characteristics—i.e., the solutions of the characteristic equations (7.11) whose initial conditions lie on this initial curve—are given by

$$x_1 = x_1(r, s), \qquad x_2 = x_2(r, s), \qquad v = v(r, s) \tag{7.12}$$

which yields a description of the solution surface of our PDE (7.10). However, we are looking for a different representation of this surface, namely, $v = v(x_1, x_2)$. To bring (7.12) to the desired form, we need to express r, s as functions of x_1, x_2 and plug them into $v(r, s)$. For this to be possible, the map $(r, s) \mapsto (x_1(r, s), x_2(r, s))$ must be invertible. It follows from the Inverse Function Theorem that invertibility of this map, at least in some neighborhood of the initial curve, is in turn guaranteed if the corresponding Jacobian matrix

$$\begin{pmatrix} (x_1)_r & (x_1)_s \\ (x_2)_r & (x_2)_s \end{pmatrix}$$

is nonsingular along the initial curve. What is the geometric significance of this latter condition? The columns of the above Jacobian matrix are the (x_1, x_2)-components of the tangent vectors to the initial curve and to a characteristic, respectively, and they must be linearly independent. In other words, when projected onto the (x_1, x_2)-plane, the characteristics and the initial curve must be *transversal* to each other. We conclude that under this transversality condition, the Cauchy problem for (7.10) is (locally) uniquely solvable, at least in principle, via the method of characteristics. To calculate the actual value of $v(x_1, x_2)$ for some specific x_1 and x_2, we need to find a characteristic that connects the initial curve to a point with these (x_1, x_2)-coordinates and then integrate the last equation in (7.11) along that characteristic segment. We will discuss this procedure in more detail later for the case of the HJB equation.

Characteristics can also be defined for the general PDE (7.9) which for $n = 2$ takes the form

$$F(x_1, x_2, v, v_{x_1}, v_{x_2}) = 0. \tag{7.13}$$

In the case of the quasi-linear PDE (7.10), we saw that the tangent vector $(a, b, c)^T$ to the solution surface does not depend on the partial derivatives v_{x_1} and v_{x_2} (which give the first two components of the normal vector $(v_{x_1}, v_{x_2}, -1)^T$ to this surface); consequently, the characteristic equations (7.11) involve x_1, x_2, and v only. For the more general PDE (7.13), this is no longer guaranteed and so five differential equations—describing the joint evolution of the variables $x_1, x_2, v, v_{x_1}, v_{x_2}$—are needed instead of three. The solution curves of these equations are called the *characteristic strips*. To simplify the notation, let us introduce the symbols

$$\xi_1 := v_{x_1}, \qquad \xi_2 := v_{x_2}.$$

Then the characteristic strips are defined by the following equations:

$$\frac{dx_1}{ds} = F_{\xi_1}, \qquad \frac{dx_2}{ds} = F_{\xi_2}, \qquad \frac{dv}{ds} = \xi_1 F_{\xi_1} + \xi_2 F_{\xi_2},$$
$$\frac{d\xi_1}{ds} = -F_{x_1} - \xi_1 F_v, \qquad \frac{d\xi_2}{ds} = -F_{x_2} - \xi_2 F_v. \qquad (7.14)$$

We do not give a complete derivation of (7.14) but note that the first three equations in (7.14) reduce to the earlier characteristic equations (7.11) in the case of the quasi-linear PDE (7.10), whereas to arrive at the fourth equation in (7.14) it is enough to write $\frac{d\xi_1}{ds} = (\xi_1)_{x_1} \frac{dx_1}{ds} + (\xi_1)_{x_2} \frac{dx_2}{ds}$, use the first two characteristic equations to rewrite this result as $(\xi_1)_{x_1} F_{\xi_1} + (\xi_1)_{x_2} F_{\xi_2}$, and then apply the identity $F_{x_1} + F_v \xi_1 + F_{\xi_1} (\xi_1)_{x_1} + F_{\xi_2} (\xi_2)_{x_1} = 0$ which follows by differentiating the PDE (7.13) with respect to x_1; the last equation in (7.14) is derived in the same manner. The last two equations in (7.14), which describe the evolution of the normal vector to the solution surface of the PDE, will play an important role in the next subsection where we turn to the HJB equation.

In the above discussion we assumed that the independent variable x lives in the plane (i.e., $n = 2$). However, it is straightforward to write down the equations of the characteristic strips for a general dimension $n \geq 2$: we must simply replace the indices $1, 2$ in (7.14) by $i \in \{1, \ldots, n\}$. In the next subsection we will work in this more general setting.

7.2.2   Canonical equations as characteristics of the HJB equation

Our goal now is to show that Hamilton's canonical equations arise as the characteristics—or, more precisely, characteristic strips—of the HJB equation. With some abuse of notation, let us write the HJB equation as

$$V_t(t, x) - H(t, x, -V_x(t, x)) = 0. \qquad (7.15)$$

Note that we did not include the control u as an argument in H; instead, we assume that the control that maximizes the Hamiltonian has already been plugged in. (This formulation also covers the calculus of variations and mechanics settings, where the velocity variable is eliminated via the Legendre transformation.) We know that we can bring the PDE (7.15) to the form (7.9) by introducing the extra state variable $x_{n+1} := t$ (cf. Section 5.3.3). We can thus write down the characteristic equations (7.14) for this particular PDE. Since the time t is present in (7.15), we can use it as the independent variable and write $\dot{x}_1$, etc. We also define the costates $p_i :=$ $-\xi_i = -v_{x_i}$, which is consistent with our earlier convention (see Section 5.2). Then the equations (7.14) immediately yield

$$\dot{x}_i = H_{p_i}, \qquad i = 1, \ldots, n$$

(plus one additional equation $\dot{x}_{n+1} = 1$ which is redundant) as well as

$$\dot{p}_i = -H_{x_i}, \qquad i = 1, \ldots, n+1$$

where we used the fact that H does not depend on V (hence the term F_v in (7.14) vanishes). Thus we have indeed recovered the familiar canonical differential equations.

Finally, there is one more equation in (7.14) which describes the evolution of v. Here it tells us that the value function satisfies

$$\dot{V} = \sum_{i=1}^{n} \xi_i \dot{x}_i + \xi_{n+1} = H - \sum_{i=1}^{n} p_i H_{p_i} \tag{7.16}$$

where we applied the HJB equation (7.15) one more time to arrive at the second equality. (Actually, it is also easy to obtain this result directly by writing $\dot{V} = V_t + \langle V_x, \dot{x} \rangle$.) It is the equation (7.16) that enables a solution of the HJB equation via the method of characteristics. We know that the Cauchy problem for the HJB equation is given by $V(t_1, x) = K(x)$, where K is the terminal cost and t_1 is the terminal time (this initial value problem involves flowing backward in time). To calculate the value $V(t, x)$ for some specific time t and state x, we must first find a point $\bar{x}$ such that $x = x(t)$ where $(x(\cdot), p(\cdot))$ is the solution of the system of canonical equations with the boundary conditions $x(t_1) = \bar{x}$, $p(t_1) = -K_x(\bar{x})$. Then, $V(t, x)$ is computed by integration along the corresponding characteristic:

$$V(t, x(t)) = K(\bar{x}) - \int_t^{t_1} \left(H(s, x(s), p(s)) - \sum_{i=1}^{n} p_i(s) H_{p_i}(s, x(s), p(s)) \right) ds.$$

Again, the fact that H does not depend on V is helpful here. Figure 7.3 illustrates the procedure.

7.3 RICCATI EQUATIONS AND INEQUALITIES IN ROBUST CONTROL

This section builds on Chapter 6, with the goal of demonstrating that the tools and results of that chapter—in particular, optimality conditions expressed in terms of algebraic Riccati equations—continue to play a central role in the context of more general control problems. The first problem studied here is a special case of the infinite-horizon LQR problem except that

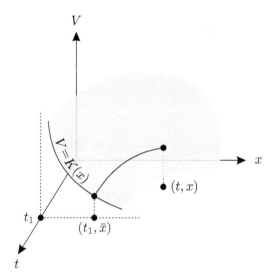

Figure 7.3: Solving the HJB equation with the help of characteristics

the matrix Q in the cost functional is no longer positive semidefinite, which leads to its interpretation as the problem of characterizing the disturbance-to-output $\mathcal{L}_2$ gain. Afterwards, we take a brief look at the $\mathcal{H}_\infty$ control problem, which has been an important benchmark problem in robust control theory.

7.3.1 $\mathcal{L}_2$ gain

We consider a linear time-invariant system

$$\dot{x} = Ax + Bu, \qquad y = Cx \qquad (7.17)$$

and the cost functional

$$J(u) = \int_{t_0}^{\infty} \left(\gamma u^T(t)u(t) - \frac{1}{\gamma} y^T(t)y(t) \right) dt \qquad (7.18)$$

with $\gamma > 0$. In this cost, the squared norms of the input and the output inside the integral are multiplied by scalar weights of the opposite signs (and we assume the two weights to have been normalized so that their product equals -1). Consequently, it is clear that the optimal cost will now be nonpositive (just set $u \equiv 0$). Nevertheless, as we will see, the form of the optimal solution is very similar to the one we saw in Section 6.2 and can be established using similar calculations. Suppose that there exists a matrix P with the following three properties:

1) $P = P^T \geq 0$.

2) P is a solution of the ARE

$$PA + A^T P + \frac{1}{\gamma} C^T C + \frac{1}{\gamma} P B B^T P = 0. \tag{7.19}$$

3) The matrix $A + \frac{1}{\gamma} B B^T P$ is Hurwitz.[2]

Then, we claim that the optimal cost is

$$V(x_0) = -x_0^T P x_0 \tag{7.20}$$

and the optimal control is the linear state feedback

$$u^*(t) = \frac{1}{\gamma} B^T P x^*(t). \tag{7.21}$$

(Notice the "wrong" signs in the formulas (7.19)–(7.21) compared to Section 6.2; this sign difference could be reconciled by working with $-P$ instead of P here.)

To prove this claim, let us define the function $\widehat{V}(x) := x^T P x$. Its derivative along solutions of the system (7.17) is $\frac{d}{dt}\widehat{V}(x(t)) = x^T(t)(PA + A^T P)x(t) + 2x^T(t)PBu(t)$, which is easily checked to be equivalent to

$$\frac{d}{dt}\widehat{V}(x(t)) = x^T(t)\left(PA + A^T P + \frac{1}{\gamma} C^T C + \frac{1}{\gamma} P B B^T P\right)x(t) \tag{7.22}$$

$$-\gamma\left(u(t) - \frac{1}{\gamma} B^T P x(t)\right)^T\left(u(t) - \frac{1}{\gamma} B^T P x(t)\right) + \gamma u^T(t)u(t) - \frac{1}{\gamma} y^T(t)y(t).$$

We now introduce the auxiliary finite-horizon cost

$$\bar{J}^{t_1}(u) := \int_{t_0}^{t_1}\left(\gamma u^T(t)u(t) - \frac{1}{\gamma} y^T(t)y(t)\right)dt - x^T(t_1)Px(t_1). \tag{7.23}$$

Using the formula (7.22) and noting that the first term on its right-hand side vanishes by (7.19), we can rewrite this cost as

$$\bar{J}^{t_1}(u) = \int_{t_0}^{t_1} \gamma\left|u(t) - \frac{1}{\gamma} B^T P x(t)\right|^2 dt + x^T(t)Px(t)\Big|_{t_0}^{t_1} - x^T(t_1)Px(t_1)$$

$$= \int_{t_0}^{t_1} \gamma\left|u(t) - \frac{1}{\gamma} B^T P x(t)\right|^2 dt - x_0^T P x_0$$

which makes it clear that (7.20) and (7.21) are the optimal cost and optimal control for the cost functional (7.23). We want to show that they are also

---

[2]A matrix is called *Hurwitz* if all its eigenvalues have negative real parts.

optimal for the original cost functional (7.18). To this end, we first note
that since $P \geq 0$, the following bound holds for all u:

$$
\begin{aligned}
J(u) &= \lim_{t_1 \to \infty} \int_{t_0}^{t_1} \left(\gamma u^T(t) u(t) - \frac{1}{\gamma} y^T(t) y(t) \right) dt \\
&\geq \lim_{t_1 \to \infty} \left(\int_{t_0}^{t_1} \left(\gamma u^T(t) u(t) - \frac{1}{\gamma} y^T(t) y(t) \right) dt - x^T(t_1) P x(t_1) \right) \\
&= \lim_{t_1 \to \infty} J^{t_1}(u) \geq \lim_{t_1 \to \infty} (-x_0^T P x_0) = -x_0^T P x_0.
\end{aligned}
$$

On the other hand, having already established that $\bar{J}^{t_1}(u^*)$ equals

$$
\int_{t_0}^{t_1} \left(\gamma (u^*)^T(t) u^*(t) - \frac{1}{\gamma} (y^*)^T(t) y^*(t) \right) dt - (x^*)^T(t_1) P x^*(t_1) = -x_0^T P x_0
$$

where $y^* := C x^*$, we can pass to the limit as $t_1 \to \infty$ in this relation. In view
of the fact that $A + \frac{1}{\gamma} B B^T P$ is a Hurwitz matrix, the closed-loop system
$\dot{x}^* = A x^* + B u^* = (A + \frac{1}{\gamma} B B^T P) x^*$ is exponentially stable. We thus obtain
$J(u^*) = -x_0^T P x_0$, and the desired result is proved.

While the formulas appearing here and in Section 6.2 are similar, the
meanings of the two problems are very different. The cost (7.18) no longer
reflects the objective of keeping both y and u small. Instead, this cost is
small when y is large relative to u. We can regard u here not as a control
that regulates the output but as a disturbance that tries to make the output
large, with the optimal input being in some sense the worst-case disturbance.
Let us try to formulate this idea in more precise terms. The fact that (7.20)
is the optimal cost for the functional (7.18) implies that the inequality

$$
-x_0^T P x_0 \leq \int_{t_0}^{\infty} \left(\gamma |u(t)|^2 - \frac{1}{\gamma} |y(t)|^2 \right) dt \tag{7.24}
$$

holds for all u, with the equality achieved by the optimal control u^*. From
now on we focus on the case when $x_0 = 0$. Specializing (7.24) to this case,
after simple manipulations we reach

$$
\sup_{u \in \mathcal{L}_2 \setminus \{0\}} \frac{\sqrt{\int_{t_0}^{\infty} |y(t)|^2 dt}}{\sqrt{\int_{t_0}^{\infty} |u(t)|^2 dt}} \leq \gamma. \tag{7.25}
$$

The fraction on the left-hand side of (7.25) is the ratio of the $\mathcal{L}_2$ norms[3] of
the input and the output, and the supremum is being taken over all nonzero
inputs with finite $\mathcal{L}_2$ norms. If we view the system (7.17), with the zero
initial condition, as an input/output operator from $\mathcal{L}_2$ to $\mathcal{L}_2$, then (7.25)

---

[3]See the definition (1.31) on page 19.

says that the induced norm of this operator does not exceed γ. This induced norm is called the $\mathcal{L}_2$ *gain* of the system.

We see that if, for a given value of γ, we can find a matrix P with the three properties listed at the beginning of this subsection, then the system's $\mathcal{L}_2$ gain is less than or equal to γ. (We do not know whether γ is actually achieved by some control; note that the optimal control (7.21) is excluded in (7.25) because it is identically 0 when $x_0 = 0$.) A converse result also holds: if the $\mathcal{L}_2$ gain is less than γ, then a matrix P with the indicated properties exists. If we sidestep the original optimal control problem and only seek sufficient conditions for the $\mathcal{L}_2$ gain to be less than or equal to γ, then it is not hard to see from our earlier derivation that the conditions on the matrix P can be relaxed. Namely, it is enough to look for a symmetric positive semidefinite solution of the algebraic Riccati *inequality*

$$PA + A^T P + \frac{1}{\gamma}C^T C + \frac{1}{\gamma}PBB^T P \leq 0. \tag{7.26}$$

The formula (7.22) then yields

$$\frac{d}{dt}\widehat{V}(x(t)) \leq \gamma |u(t)|^2 - \frac{1}{\gamma}|y(t)|^2.$$

Integrating both sides from t_0 to an arbitrary time T, rearranging terms, and using the definition of $\widehat{V}$ and the fact that $P \geq 0$, we have

$$\frac{1}{\gamma}\int_{t_0}^{T}|y(t)|^2 dt \leq \gamma\int_{t_0}^{T}|u(t)|^2 dt + \widehat{V}(0) - \widehat{V}(x(t)) \leq \gamma\int_{t_0}^{T}|u(t)|^2 dt$$

and in the limit as $T \to \infty$ we again arrive at (7.25).

In the frequency domain, the system (7.17) is characterized by the transfer matrix $G(s) = C(Is - A)^{-1}B$. Using Parseval's theorem, it can be shown that the $\mathcal{L}_2$ gain equals the largest singular value of $G(j\omega)$ supremized over all frequencies $\omega \in \mathbb{R}$; for systems with scalar inputs and outputs, this is just $\sup_{\omega \in \mathbb{R}}|g(j\omega)|$ where g is the transfer function. In view of this fact, the $\mathcal{L}_2$ gain is also called the $\mathcal{H}_\infty$ *norm*.

7.3.2 $\mathcal{H}_\infty$ control problem

We saw in the previous subsection how u can play the role of a disturbance input, in contrast with the standard LQR setting of Chapter 6 where, as in the rest of this book, it plays the role of a control input. Now, let us make the situation more interesting by allowing both types of inputs to be present in the system, the task of the control being to stabilize the system and attenuate the unknown disturbance in some sense. Such control problems fall into the general framework of *robust control theory*, which deals with control design methods for providing a desired behavior in the presence of uncertainty. (We can also think of the control and the disturbance as two opposing players in a differential game.) In the specific problem considered in this subsection, the level of disturbance attenuation will be measured by the $\mathcal{L}_2$ gain (or $\mathcal{H}_\infty$ norm) of the closed-loop system.

The $\mathcal{H}_\infty$ control problem concerns itself with a system of the form

$$\dot{x} = Ax + Bu + Dw, \qquad y = Cx, \qquad z = Ex$$

where u is the control input, w is the disturbance input, y is the measured output (the quantity available for feedback), and z is the controlled output (the quantity to be regulated); for simplicity, we assume here that there are no feedthrough terms from u and w to y and z. The corresponding feedback control diagram is shown in Figure 7.4. We will first consider the simpler case of *state feedback*, obtained by setting $C := I$ so that $y = x$. In this case, we also restrict our attention to controllers that take the static linear state feedback form $u = Kx$.

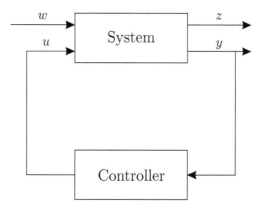

Figure 7.4: H_∞ control problem setting

The control objective is to stabilize the internal dynamics and attenuate w in the $\mathcal{H}_\infty$ sense. More specifically, we want to design the feedback gain matrix K so that the following two properties hold:

1) The closed-loop system matrix $A_{\mathrm{cl}} := A + BK$ is Hurwitz.

2) The $\mathcal{L}_2$ gain of the closed-loop system from w to z (or, what is the same, the $\mathcal{H}_\infty$ norm of the closed-loop transfer matrix $G(s) = E(Is - A_{\mathrm{cl}})^{-1}D$) does not exceed a prespecified value $\gamma > 0$.

Note that this is not an optimal control problem because we are not asking to minimize the gain γ (although ideally of course we would like γ to be as small as possible). Controls solving problems of this kind are known as *suboptimal*.

It follows from the results of the previous subsection—applied with the change of notation from A, B, C, u, y to $A_{\mathrm{cl}}, D, E, w, z$, respectively—that to have property 2 (the bound on the $\mathcal{L}_2$ gain) it is sufficient to find a positive semidefinite solution of the Riccati inequality

$$PA_{\mathrm{cl}} + A_{\mathrm{cl}}^T P + \frac{1}{\gamma} E^T E + \frac{1}{\gamma} PDD^T P \leq 0. \tag{7.27}$$

To also guarantee property 1 (internal stability of the closed-loop system) requires slightly stronger conditions: P needs to be positive definite and the inequality in (7.27) needs to be strict. The latter condition can be encoded via the Riccati equation

$$PA_{\mathrm{cl}} + A_{\mathrm{cl}}^T P + \frac{1}{\gamma} E^T E + \frac{1}{\gamma} PDD^T P + \varepsilon Q = 0 \tag{7.28}$$

where $Q = Q^T > 0$ and $\varepsilon > 0$. (This implies $PA_{\mathrm{cl}} + A_{\mathrm{cl}}^T P < 0$ which is the well-known Lyapunov condition for A_{cl} to be a Hurwitz matrix.) Next, we need to convert (7.28) into a condition that is verifiable in terms of the original open-loop system data. Introducing a matrix $R = R^T > 0$ as another design parameter (in addition to Q and ε), suppose that there exists a solution $P > 0$ to the Riccati equation

$$PA + A^T P + \frac{1}{\gamma} E^T E + \frac{1}{\gamma} PDD^T P - \frac{1}{\varepsilon} PBR^{-1}B^T P + \varepsilon Q = 0. \tag{7.29}$$

Then, if we let $K := -\frac{1}{2\varepsilon} R^{-1} B^T P$, a straightforward calculation shows that the feedback law $u = Kx$ enforces (7.28) and thus achieves both of our control objectives. Conversely, it can be shown that if the system is stabilizable with an $\mathcal{L}_2$ gain less than γ, then (7.29) is solvable for $P > 0$.

The general case—when the full state is not measured and the controller is a dynamic output feedback—is more interesting and more complicated. Without going into details, we mention that a complete solution to this problem, in the form of necessary and sufficient conditions for the existence of a controller achieving an $\mathcal{L}_2$ gain less than γ, is available and consists of the following ingredients:

1) Finding a solution P_1 of a Riccati equation from the state feedback case.

2) Finding a solution P_2 of another Riccati equation obtained from the first one by the substitutions $B \to C$ and $D \leftrightarrow E$.

3) Checking that the largest singular value of the product $P_1 P_2$ is less than γ^2.

These elegant conditions in terms of two coupled Riccati equations yield a controller that can be interpreted as a state feedback law combined with an estimator (observer).

7.3.3 Riccati inequalities and LMIs

As we have seen, a representative example of Riccati matrix inequalities encountered in robust control is

$$PA + A^T P + C^T C + PBB^T P \leq 0 \qquad (7.30)$$

(this is (7.26) with $\gamma = 1$). Similar Riccati inequalities arise in the context of several other control problems. The inequality (7.30) is quadratic in the matrix variable P. However, it can be shown to be equivalent to the *linear matrix inequality* (LMI)

$$\begin{pmatrix} PA + A^T P + C^T C & PB \\ B^T P & -I \end{pmatrix} \leq 0. \qquad (7.31)$$

The matrix on the left-hand side of (7.31) is called the *Schur complement* of the matrix on the left-hand side of (7.30). The constraint $P \geq 0$ can also be naturally incorporated into this LMI.

Being convex feasibility problems, LMIs can be efficiently solved by known numerical algorithms from convex optimization. Several dedicated software packages exist for solving LMIs. This makes conditions expressed in terms of LMIs attractive from the computational point of view.

7.4 MAXIMUM PRINCIPLE FOR HYBRID CONTROL SYSTEMS

Hybrid systems are systems whose dynamics involve a combination of continuous evolution and discrete transitions. More specifically, in this section we consider hybrid systems described by a finite collection of control systems and a finite sequence of times (called *switching times*) which partition the time interval into subintervals. On each subinterval, the state of the system flows in accordance with one of the systems from a given collection; at a switching time, the state experiences an instantaneous jump, and another system from the collection is selected for the next subinterval. While providing a much richer modeling framework than the continuous control systems considered elsewhere in this book, hybrid systems violate the assumptions under which we developed the maximum principle in Chapter 4. In this section, we discuss a suitably extended version of the maximum principle which applies to hybrid systems.

7.4.1 Hybrid optimal control problem

We begin by defining the class of hybrid control systems and associated optimal control problems that we want to study. The first ingredient of our hybrid control system is a collection of (time-invariant) control systems

$$\dot{x} = f_q(x, u), \qquad q \in Q \tag{7.32}$$

where Q is a finite index set; for simplicity, we assume that all the systems in the collection (7.32) share the same state space $\mathbb{R}^n$ and control set $U \subset \mathbb{R}^m$. The second ingredient is a collection of *switching sets* $S_{q,q'} \subset \mathbb{R}^{2n}$, one for each pair $(q, q') \in Q \times Q$. A function $x : [t_0, t_f] \to \mathbb{R}^n$ is an admissible trajectory of our hybrid system corresponding to a control $u : [t_0, t_f] \to U$ if there exist time instants

$$t_0 < t_1 < \cdots < t_k < t_{k+1} := t_f$$

and indices $q_0, q_1, \ldots, q_k \in Q$ such that $x(\cdot)$ satisfies

$$\dot{x}(t) = f_{q_i}(x(t), u(t)) \qquad \forall t \in (t_i, t_{i+1}), \; i = 0, 1, \ldots, k \tag{7.33}$$

and

$$\begin{pmatrix} x(t_i^-) \\ x(t_i^+) \end{pmatrix} \in S_{q_{i-1}, q_i}, \qquad i = 1, \ldots, k.$$

Here $x(t_i^-)$ and $x(t_i^+)$ are the values of x right before and right after t_i, respectively, and the value $x(t_i)$ is taken to be equal to one of these one-sided limits, depending on the desired convention. At each possible discontinuity t_i, a *discrete transition* (or *switching event*) is said to occur. The function $q : [t_0, t_f] \to Q$ defined by $q(t) := q_i$ for $t \in [t_i, t_{i+1})$ describes the evolution of q along the trajectory; q is often called the *discrete state* of the hybrid system. We can use it to rewrite (7.33) more concisely as $\dot{x}(t) = f_{q(t)}(x(t), u(t))$ for $t \neq t_i$, $i = 1, \ldots, k$.

We consider cost functionals of the form

$$J(u, \{t_i\}, \{q_i\}) := \sum_{i=0}^{k} \int_{t_i}^{t_{i+1}} L_{q_i}(x(t), u(t)) dt + \sum_{i=1}^{k} \Phi_{q_{i-1}, q_i}(x(t_i^-), x(t_i^+)) \tag{7.34}$$

where $L_q : \mathbb{R}^n \times U \to \mathbb{R}$ is the usual running cost and $\Phi_{q,q'} : \mathbb{R}^n \times \mathbb{R}^n \to \mathbb{R}$ is the *switching cost*, for $q, q' \in Q$. For simplicity, we do not include a terminal cost (this is no loss of generality, as terminal cost can be easily incorporated into the above running-plus-switching cost along the lines of Section 3.3.2). We also introduce an *endpoint constraint*, characterized by a set $E_{q,q'} \subset \mathbb{R}^{2n}$ for each pair $(q, q') \in Q \times Q$, according to which the trajectory $x(\cdot)$ must satisfy

$$\begin{pmatrix} x(t_0) \\ x(t_f) \end{pmatrix} \in E_{q_0, q_k}. \tag{7.35}$$

(Here E_{q_0, q_k} plays the same role as S_2 at the end of Section 4.3.1.) Then, the hybrid optimal control problem consists in finding a control that minimizes the cost (7.34) subject to the endpoint constraint (7.35). We emphasize that the choice of a control u is accompanied by the choice of two finite sequences $\{t_i\}$ and $\{q_i\}$, to which we henceforth refer as the *time sequence* and *switching sequence*, respectively.

7.4.2 Hybrid maximum principle

The maximum principle that we are about to state provides necessary conditions for a trajectory $x^*(\cdot)$ of the hybrid control system corresponding to a control $u^*(\cdot)$, a time sequence $\{t_i\}$, and a switching sequence $\{q_i\}$ to be locally optimal over trajectories $x(\cdot)$ with the same switching sequence $\{q_i\}$ and such that x is close to x^* on each subinterval (t_i, t_{i+1}). Most of the statements of this hybrid maximum principle are more or less familiar to us from Chapter 4. We proceed with the understanding that suitable technical assumptions are in place so that all derivatives, tangent spaces, and other objects appearing below are well defined.

Define the family of Hamiltonians

$$H_q(x, u, p, p_0) := \langle p, f_q(x, u)\rangle + p_0 L_q(x, u), \qquad q \in Q.$$

The abnormal multiplier must satisfy $p_0^* \leq 0$ as usual. The costate $p^*(\cdot)$ is allowed to be discontinuous at the switching times t_i of x^*, while between these times it must satisfy the adjoint equation

$$\dot{p}^*(t) = -\left.(H_{q_i})_x\right|_*(t) \qquad \forall\, t \in (t_i, t_{i+1}),\ i = 0, 1, \ldots, k.$$

The *transversality condition* says that the vector $\begin{pmatrix} p^*(t_0) \\ -p^*(t_f) \end{pmatrix}$ must be orthogonal to the tangent space to the endpoint constraint set E_{q_0,q_k} at $\begin{pmatrix} x^*(t_0) \\ x^*(t_f) \end{pmatrix}$, which we write as

$$\begin{pmatrix} p^*(t_0) \\ -p^*(t_f) \end{pmatrix} \perp T_{\begin{pmatrix} x^*(t_0) \\ x^*(t_f) \end{pmatrix}} E_{q_0,q_k}. \tag{7.36}$$

At the switching times, there are also *switching conditions* saying that for $i = 1, \ldots, k$ we must have

$$\begin{pmatrix} -p^*(t_i^-) \\ p^*(t_i^+) \end{pmatrix} + p_0^* \nabla \Phi_{q_{i-1},q_i}(x^*(t_i^-), x^*(t_i^+)) \perp T_{\begin{pmatrix} x^*(t_i^-) \\ x^*(t_i^+) \end{pmatrix}} S_{q_{i-1},q_i}. \tag{7.37}$$

The *nontriviality condition* says that either $p_0^* \neq 0$ or $p^* \not\equiv 0$. The *Hamiltonian maximization condition* must hold in the usual sense for each H_{q_i} on the corresponding interval (t_i, t_{i+1}). Moreover, the Hamiltonian remains constant along the optimal trajectory (in particular, its value is not affected by the switching events). Finally, for free-time problems the Hamiltonian is 0.

Note that the transversality condition (7.36) is completely analogous to the transversality condition (4.46) for the case of initial sets discussed at the end of Section 4.3.1. As for the switching conditions (7.37), the intuition behind them is similar and can be understood as follows. Consider the continuous portions of x^* which correspond to the subintervals (t_i, t_{i+1}), $i = 0, \ldots, k$. Reparameterize the time individually for each of them so that their domains are all mapped onto the same interval, say, $[s_0, s_f]$. This allows us to "stack" them all together, i.e., treat them as if they evolve simultaneously. Then, the transversality condition (7.36) and the switching conditions (7.37) become one aggregate transversality condition induced by the endpoint constraint and the switching sets. The appearance of the gradient of the switching cost in this transversality condition is also not surprising because the switching cost becomes a combination of terminal and initial cost (see Section 4.3.1 for a discussion of transversality conditions for problems with terminal cost).

7.4.3 Example: light reflection

To illustrate the hybrid maximum principle, we apply it to the familiar light reflection example from Section 2.1.2. We model the propagation of a light ray through the n-dimensional space via the control system

$$\dot{x} = c(x)u, \qquad |u| = 1 \tag{7.38}$$

where $x \in \mathbb{R}^n$, $c : \mathbb{R}^n \to (0, \infty)$ is a $\mathcal{C}^1$ function that determines the (varying) speed of light, and u taking values on the unit sphere in $\mathbb{R}^n$ defines the direction of motion. We assume that the reflecting surface is a hyperplane, and without loss of generality we take it to be $S := \{x : x_n = 0\}$. The initial point and the final point are assumed to lie in the same open half-space relative to S.

We seek to derive a necessary condition for a trajectory x^* that starts at a given initial point x_0 at time t_0, gets reflected off S at some time t_1, and arrives at a given final point x_f at time t_f to be locally time-optimal with respect to trajectories that hit S at nearby points. This optimal control problem is not inherently hybrid, since (7.38) is a standard control system and it is capable of producing reflected trajectories. However, the classical formulation of the maximum principle does not allow us to incorporate the fact that we are only interested in trajectories that hit S along the way. With the hybrid formulation, it is easy to do so by considering a hybrid system with a single discrete state location q (i.e., $Q = \{q\}$) and the switching set $S_{q,q} := \left\{ \begin{pmatrix} x \\ x' \end{pmatrix} : x = x' \in S \right\}$. In this hybrid system, discrete transitions occur when the trajectory hits S, but the underlying control system (7.38) does not change and the trajectory remains continuous. The switching sequence associated with our candidate optimal trajectory x^* is $\{q, q\}$, and the hybrid maximum principle captures local optimality over nearby trajectories with the same switching sequence—which is precisely what we want.

Applying the hybrid maximum principle to this problem entails just a few straightforward computations. The Hamiltonian is $H = \langle p, c(x)u \rangle + p_0$. The Hamiltonian maximization condition gives $u^* = p^*/|p^*|$ and $H|_* = c(x^*)|p^*| + p_0^*$. Since the final time is free, we have $H|_* \equiv 0$ which in view of the nontriviality condition implies that $p_0^* \neq 0$ and $p^*(t) \neq 0$ for all t. Normalizing them so that $p_0^* = -1$, we obtain $c(x^*)|p^*| \equiv 1$ hence $u^* = p^* c(x^*)$. Both the costate p^* and the optimal control u^* are continuous except possibly at the switching time t_1. Since there is no switching cost, the switching condition tells us that the vector $\begin{pmatrix} -p^*(t_1^-) \\ p^*(t_1^+) \end{pmatrix}$ must be orthogonal to the tangent space to $S_{q,q}$ at $\begin{pmatrix} x^*(t_1) \\ x^*(t_1) \end{pmatrix}$. This tangent space is $S_{q,q}$ itself, and vectors in it have the form $(x_1, \ldots, x_{n-1}, 0, x_1, \ldots, x_{n-1}, 0)^T$. It follows that $p_i^*(t_1^+) = p_i^*(t_1^-)$ for $i = 1, \ldots, n-1$; in other words, $p_1^*, \ldots, p_{n-1}^*$ are continuous at t_1, hence so are $u_1^*, \ldots, u_{n-1}^*$. Only the last component of u^* can be discontinuous at t_1. But since $|u^*| = \sqrt{(u_1^*)^2 + \cdots + (u_n^*)^2}$ is to remain equal to 1, it must be that $u_n^*(t_1^+) = \pm u_n^*(t_1^-)$, i.e., u_n^* either stays continuous or flips its sign at t_1. Of these two options, only the latter is possible because the light ray cannot pass through S. We conclude that the velocity vectors before and after the reflection differ only in the component orthogonal to the reflection surface, and the difference is only in the minus sign. We have of course recovered the well-known law of reflection.

7.5   NOTES AND REFERENCES FOR CHAPTER 7

Our primary reference on manifolds and some related mathematical facts was [Arn89] (see in particular Sections 18, 34, and 37 of that book). Most of this material is also presented in [Jur96] where the maximum principle on manifolds is discussed too. Two other good sources of basic information about manifolds, on which we occasionally relied in Section 7.1, are [Arn92, Chapter 5] and [Isi95, Appendix A]; for further reading on this subject we recommend [Boo03] or [War83]. An in-depth treatment of the maximum principle on manifolds can be found in [Sus97] as well as in [AS04, Chapter 12]; see also [Cha11].

A general source of information on PDEs and the method of characteristics is [Gar86]; in our presentation in Section 7.2 we drew upon Chapter 2 of that book (which also has a section on the Hamilton-Jacobi theory). References dealing specifically with the connection between the HJB equation and the canonical equations via characteristics (without going into the details of the general PDE theory) are [YZ99, Chapter 5] and [BP07, Chapter 7].

All the ingredients of our treatment of the $\mathcal{L}_2$ gain problem are contained in [Bro70, Sections 23 and 25] and [Kha02, Section 5.3]. Our derivation of the state feedback $\mathcal{H}_\infty$ controller follows the paper [Pet87], whereas the general output feedback $\mathcal{H}_\infty$ controller is presented in the paper [DGKF89] as well as the book [ZDG96]. A good reference on LMIs, methods for solving them, and their role in system and control theory is [BGFB94], while [HJ85] supplies relevant technical details on Schur complements. Although we only considered linear systems for simplicity, most of the concepts described in Section 7.3 can be as naturally developed for nonlinear systems (with Hamilton-Jacobi partial differential inequalities replacing Riccati matrix inequalities); besides the already mentioned text [Kha02], this topic is discussed in much greater detail in [vdS96].

Section 7.4 is based on the paper [Sus99]. Related work is reported in the papers [GP05] and [DK08], the latter of which contains the "stacking" argument to which we alluded at the end of Section 7.4.2. A general reference on hybrid systems is [vdSS00]. Optimal control of hybrid systems is an active research area; see, e.g., [BWEV05] and the references therein.

Bibliography

[AF66]   M. Athans and P. L. Falb. *Optimal Control*. McGraw Hill, New York, 1966. Reprinted by Dover in 2006.

[AM90]   B.D.O. Anderson and J. B. Moore. *Optimal Control: Linear Quadratic Methods*. Prentice Hall, New Jersey, 1990. Reprinted by Dover in 2007.

[Arn89]   V. I. Arnold. *Mathematical Methods of Classical Mechanics*. Springer, New York, 2nd edition, 1989.

[Arn92]   V. I. Arnold. *Ordinary Differential Equations*. Springer, Berlin, 3rd edition, 1992.

[AS04]   A. A. Agrachev and Yu. L. Sachkov. *Control Theory from the Geometric Viewpoint*. Springer, Berlin, 2004.

[BCD97]   M. Bardi and I. Capuzzo-Dolcetta. *Optimal Control and Viscosity Solutions of Hamilton-Jacobi-Bellman Equations*. Birkhäuser, Boston, 1997.

[Bel57]   R. Bellman. *Dynamic Programming*. Princeton University Press, 1957.

[Ber99]   D. P. Bertsekas. *Nonlinear Programming*. Athena Scientific, Belmont, MA, 2nd edition, 1999.

[BGFB94]   S. Boyd, L. El Ghaoui, E. Feron, and V. Balakrishnan. *Linear Matrix Inequalities in System and Control Theory*, volume 15 of *SIAM Studies in Applied Mathematics*. SIAM, Philadelphia, 1994.

[Bli30]   G. A. Bliss. On the problem of Lagrange in the calculus of variations. *Amer. J. Math.*, 52:673–744, 1930.

[Blo03]   A. M. Bloch. *Nonholonomic Mechanics and Control*. Springer, New York, 2003.

[BM91]   U. Brechtken-Manderscheid. *Introduction to the Calculus of Variations*. Chapman & Hall, London, 1991.

[Bol78]   V. G. Boltyanskii. *Optimal Control of Discrete Systems*. Wiley, New York, 1978.

[Boo03]   W. M. Boothby. *An Introduction to Differentiable Manifolds and Riemannian Geometry*. Academic Press, New York, revised 2nd edition, 2003.

[BP04]   U. Boscain and B. Piccoli. *Optimal Syntheses for Control Systems on 2-D Manifolds*. Springer, New York, 2004.

[BP07]   A. Bressan and B. Piccoli. *Introduction to the Mathematical Theory of Control*. American Institute of Mathematical Sciences, 2007.

[Bre85]   A. Bressan. A high order test for optimality of bang-bang controls. *SIAM J. Control Optim.*, 23:38–48, 1985.

[Bro70]   R. W. Brockett. *Finite Dimensional Linear Systems*. Wiley, New York, 1970.

[Bry96]   A. E. Bryson Jr. Optimal control—1950 to 1985. *IEEE Control Systems Magazine*, 16:26–33, 1996.

[BV04]   S. Boyd and L. Vandenberghe. *Convex Optimization*. Cambridge University Press, 2004.

[BWEV05]   M. Boccadoro, Y. Wardi, M. Egerstedt, and E. Verriest. Optimal control of switching surfaces in hybrid dynamical systems. *Discrete Event Dyn. Syst.*, 15:433–448, 2005.

[CEHS87]   G. S. Christensen, M. E. El-Hawary, and S. A. Soliman. *Optimal Control Applications in Electric Power Systems*. Plenum Press, New York, 1987.

[Ces83]   L. Cesari. *Optimization—Theory and Applications*. Springer, New York, 1983.

[Cha11]   D. E. Chang. A simple proof of the Pontryagin maximum principle on manifolds. *Automatica*, 47:630–633, 2011.

[CL83]   M. G. Crandall and P. L. Lions. Viscosity solutions of Hamilton-Jacobi equations. *Trans. Amer. Math. Soc.*, 277:1–42, 1983.

[Cla89]   F. H. Clarke. *Methods of Dynamic and Nonsmooth Optimization*. SIAM, Philadelphia, 1989.

[Cla10]   C. W. Clark. *Mathematical Bioeconomics: The Mathematics of Conservation*. Wiley, New York, 3rd edition, 2010.

[DGKF89]   J. C. Doyle, K. Glover, P. P. Khargonekar, and B. A. Francis. State-space solutions to standard $\mathcal{H}_2$ and $\mathcal{H}_\infty$ control problems. *IEEE Trans. Automat. Control*, 34:831–847, 1989.

[DK08]   A. V. Dmitruk and A. M. Kaganovich. The Hybrid Maximum Principle is a consequence of Pontryagin Maximum Principle. *Systems Control Lett.*, 57:964–970, 2008.

[dlF00]   A. de la Fuente. *Mathematical Methods and Models for Economists*. Cambridge University Press, 2000.

[DM70]   P. Dyer and S. R. McReynolds. *The Computation and Theory of Optimal Control*. Academic Press, New York, 1970.

[Fil88]   A. F. Filippov. *Differential Equations with Discontinuous Right-hand Sides*. Kluwer, Dordrecht, 1988.

[FLS63]   R. P. Feynman, R. B. Leighton, and M. Sands. *The Feynman Lectures on Physics*. Addison-Wesley, Reading, MA, 1963.

[Ful85]   A. T. Fuller. Minimization of various performance indices for a system with bounded control. *Int. J. Control*, 41:1–37, 1985.

[Gar86]   P. R. Garabedian. *Partial Differential Equations*. Chelsea Pub. Co., New York, 2nd edition, 1986.

[GF63]   I. M. Gelfand and S. V. Fomin. *Calculus of Variations*. Prentice Hall, New Jersey, 1963. Reprinted by Dover in 2000.

[Gol80]   H. H. Goldstine. *A History of the Calculus of Variations from the 17th through the 19th Century*. Springer, New York, 1980.

[GP05]   M. Garavello and B. Piccoli. Hybrid necessary principle. *SIAM J. Control Optim.*, 43:1867–1887, 2005.

[GS07]   R. Goebel and M. Subbotin. Continuous time linear quadratic regulator with control constraints via convex duality. *IEEE Trans. Automat. Control*, 52:886–892, 2007.

[Hes09]   J. P. Hespanha. *Linear Systems Theory*. Princeton University Press, 2009.

[HJ85]   R. A. Horn and C. R. Johnson. *Matrix Analysis*. Cambridge University Press, 1985.

[Isi95]   A. Isidori. *Nonlinear Control Systems*. Springer, Berlin, 3rd edition, 1995.

[Jur96]   V. Jurdjevic. *Geometric Control Theory*. Cambridge University Press, 1996.

[Kal60]   R. E. Kalman. Contributions to the theory of optimal control. *Bol. Soc. Mat. Mexicana*, 5:102–119, 1960. Reprinted in *Control Theory: Twenty-Five Seminal Papers (1931–1981)*, T. Basar, editor, IEEE Press, New York, 2001, pages 149–166.

[Kha02]   H. K. Khalil. *Nonlinear Systems*. Prentice Hall, New Jersey, 3rd edition, 2002.

[Kno81]   G. Knowles. *An Introduction to Applied Optimal Control*. Academic Press, New York, 1981.

[Kre77]   A. J. Krener. The high order maximal principle and its applications to singular extremals. *SIAM J. Control Optim.*, 15:256–293, 1977.

[KS72]   H. Kwakernaak and R. Sivan. *Linear Optimal Control Systems*. Wiley, New York, 1972.

[Lei81]   G. Leitmann. *The Calculus of Variations and Optimal Control: An Introduction*. Plenum Press, New York, 1981.

[LL50]   M. A. Lavrentiev and L. A. Lusternik. *A Course in the Calculus of Variations*. Moscow, 2nd edition, 1950. In Russian.

[LM67]   E. B. Lee and L. Markus. *Foundations of Optimal Control Theory*. Wiley, New York, 1967.

[LSW96]   Y. Lin, E. D. Sontag, and Y. Wang. A smooth converse Lyapunov theorem for robust stability. *SIAM J. Control Optim.*, 34:124–160, 1996.

[Lue69]   D. G. Luenberger. *Optimization by Vector Space Methods*. Wiley, New York, 1969.

[Lue84]   D. G. Luenberger. *Linear and Nonlinear Programming*. Addison-Wesley, Reading, MA, 2nd edition, 1984.

[Mac05]   C. R. MacCluer. *Calculus of Variations*. Prentice Hall, New Jersey, 2005.

[McS39]   E. J. McShane. On multipliers for Lagrange problems. *Amer. J. Math.*, 61:809–819, 1939.

[MO98]   A. A. Milyutin and N. P. Osmolovskii. *Calculus of Variations and Optimal Control*. American Mathematical Society, Providence, RI, 1998.

[Neu03]   M. G. Neubert. Marine reserves and optimal harvesting. *Ecology Letters*, 6:843–849, 2003.

[NRV84]   Z. Nahorski, H. F. Ravn, and R.V.V. Vidal. The discrete-time maximum principle: a survey and some new results. *Int. J. Control*, 40:533–554, 1984.

[PB94]   H. J. Pesch and R. Bulirsch. The maximum principle, Bellman's equation, and Carathéodory's work. *J. Optim. Theory Appl.*, 80:199–225, 1994.

[PBGM62] L. S. Pontryagin, V. G. Boltyanskii, R. V. Gamkrelidze, and E. F. Mishchenko. *The Mathematical Theory of Optimal Processes*. Interscience, New York, 1962.

[Pet87]   I. R. Petersen. Disturbance attenuation and H^∞ optimization: a design method based on the algebraic Riccati equation. *IEEE Trans. Automat. Control*, 32:427–429, 1987.

[PS00]   B. Piccoli and H. J. Sussmann. Regular synthesis and sufficiency conditions for optimality. *SIAM J. Control Optim.*, 39:359–410, 2000.

[Roc74]   R. T. Rockafellar. *Conjugate Duality and Optimization*. SIAM, Philadelphia, 1974.

[Rud76]   W. Rudin. *Principles of Mathematical Analysis*. McGraw Hill, New York, 3rd edition, 1976.

[RW00]   R. T. Rockafellar and P. R. Wolenski. Convexity in Hamilton–Jacobi theory I: dynamics and duality. *SIAM J. Control Optim.*, 39:1323–1350, 2000.

[Rya87]   E. P. Ryan. Feedback solution of a class of optimal bilinear control problems. *Int. J. Control*, 45:1035–1041, 1987.

[Son98]   E. D. Sontag. *Mathematical Control Theory: Deterministic Finite Dimensional Systems*. Springer, New York, 2nd edition, 1998.

[ST05]   S. P. Sethi and G. L. Thompson. *Optimal Control Theory: Applications to Management Science and Economics*. Springer, New York, 2nd edition, 2005.

[Sus79]   H. J. Sussmann. A bang-bang theorem with bounds on the number of switchings. *SIAM J. Control Optim.*, 17:629–651, 1979.

[Sus83]   H. J. Sussmann. Lie brackets, real analyticity and geometric control. In R. W. Brockett, R. S. Millman, and H. J. Sussmann, editors, *Differential Geometric Control Theory*, pages 1–116. Birkhäuser, Boston, 1983.

[Sus97]   H. J. Sussmann. An introduction to the coordinate-free maxi-
            mum principle. In B. Jakubczyk and W. Respondek, editors, *Ge-
            ometry of Feedback and Optimal Control*, pages 463–557. Marcel
            Dekker, New York, 1997.

[Sus99]   H. J. Sussmann. A maximum principle for hybrid optimal con-
            trol problems. In *Proc. 38th IEEE Conf. on Decision and Con-
            trol*, pages 425–430, 1999.

[Sus00]   H. J. Sussmann. Handouts for the course taught at
            the Weizmann Institute of Science, 2000. Available at
            http://www.math.rutgers.edu/∼sussmann.

[Sus07]   H. J. Sussmann. Set separation, approximating multicones, and
            the Lipschitz maximum principle. *J. Differential Equations*,
            243:448–488, 2007.

[Sut75]   W. A. Sutherland. *Introduction to Metric and Topological
            Spaces*. Oxford University Press, 1975.

[SW77]   A. P. Sage and C. C. White. *Optimum Systems Control*. Prentice
            Hall, New Jersey, 2nd edition, 1977.

[SW97]   H. J. Sussmann and J. C. Willems. 300 years of optimal con-
            trol: from the brachystochrone to the maximum principle. *IEEE
            Control Systems Magazine*, 17:32–44, 1997.

[Swa84]   G. W. Swan. *Applications of Optimal Control Theory in
            Biomedicine*. Marcel Dekker, New York, 1984.

[vdS96]   A. van der Schaft. L_2-*Gain and Passivity Techniques in Non-
            linear Control*. Springer, London, 1996.

[vdSS00]   A. van der Schaft and H. Schumacher. *An Introduction to Hybrid
            Dynamical Systems*. Springer, London, 2000.

[Vin00]   R. Vinter. *Optimal Control*. Birkhäuser, Boston, 2000.

[War83]   F. W. Warner. *Foundations of Differentiable Manifolds and Lie
            Groups*. Springer, New York, 1983.

[You80]   L. C. Young. *Lectures on the Calculus of Variations and Optimal
            Control Theory*. Chelsea Pub. Co., New York, 2nd edition, 1980.

[YZ99]   J. Yong and X. Y. Zhou. *Stochastic Controls: Hamiltonian Sys-
            tems and HJB Equations*. Springer, New York, 1999.

[ZDG96]   K. Zhou, J. C. Doyle, and K. Glover. *Robust and Optimal Con-
            trol*. Prentice Hall, New Jersey, 1996.

Index